The Human Role in the Age of AI, Vol. 1: The Horizon

Guide for Every Generation Facing the Next 25 Years

A Four-Book Guide to Helping Every Generation Thrive

Dr. Harry Wong

Copyright

This book is a work of nonfiction. It reflects the author's personal analysis, professional experience, and interpretation of the social, educational, economic, and human questions raised by artificial intelligence.

The information in this book is provided for general educational and reflective purposes. It is not intended as legal, financial, medical, or professional advice.

Published by Dr. Harry Wong
First edition: 2026
Series: The Human Role in the Age of AI — A Four-Book Guide to Helping Every Generation Thrive
ISBN: 979-8-9961388-0-7 (paperback)
ISBN: 979-8-9961388-1-4 (hardcover)

Cover design: Harry Wong, with AI-assisted design support

Interior formatting: Harry Wong, with AI-assisted formatting support

This book was written and directed by the author, with AI-assisted support for editing, formatting, image-prompt development, and manuscript refinement. The author reviewed and approved the final content.

Dedication

To every generation facing the age of AI — the young, the working, the retired, and the guiding — and to all who still believe that technology should serve human dignity, wisdom, and meaning.

How to Read This Series

This book stands alone as Volume 1, the wide-angle map of the AI age. It is part of a four-volume guide designed to help families, professionals, students, and communities navigate the age of AI with clarity and courage. While each volume stands on its own and can be read in any order, the series is curated as a **Complete Family Collection** to provide a shared roadmap for everyone in your life.

The series is built to bridge the generational divide:

- The Human Role in the Age of AI, Vol. 1: The Horizon: Guide for Every Generation Facing the Next 25 Years
- The Human Role in the Age of AI, Vol. 2: The Middle Generation: How Professionals Can Adapt, Lead, and Support Their Families
- The Human Role in the Age of AI, Vol. 3: The Young Generation: How Students Can Learn, Think, and Build a Meaningful Life
- The Human Role in the Age of AI, Vol. 4: The Guiding Generation: How Parents, Grandparents, Teachers, and Mentors Can Help the Young Thrive

Why the Collection Matters

The age of AI is not only a workplace story; it is a family story. This series is written to provide a common language that allows professionals to guide their children, parents, spouses, and community members toward the future with confidence. Begin with the volume most relevant to your current responsibility, and consider sharing the complete four-volume collection to help every generation in your family navigate this transition together.

Acknowledgements

I am deeply grateful to my family, especially my wife, whose love, patience, and encouragement made this work possible. This book grew out of many years of thinking about artificial intelligence, teaching, business, music, community conversations, and family life. But at its heart, it is a book about people: the young, the working, the retired, and the adults who guide the next generation.

My children and grandchildren have helped me see the AI age not only as a technological question, but as a family and generational question. Watching the young learn, ask questions, lose and regain curiosity, and face an uncertain future has strengthened my conviction that artificial intelligence must always be discussed in human terms.

I am also grateful to the students, professionals, employees, customers, colleagues, older adults, friends, and community members whose questions and concerns helped shape this book. Their hopes, worries, curiosity, and practical experiences reminded me again and again that the future of AI is not only about machines. It is about dignity, work, education, family, wisdom, responsibility, and the kind of human future we choose to build.

Finally, I thank every reader who approaches this subject not with fear or blind excitement, but with the desire to understand how powerful technology can serve human life rather than diminish it. This book is written in that spirit.

Contents

Foreword

I first met Harry Wong in the early 1990s, when I was a product manager for Lotus Notes and the PC software industry was just beginning to grapple with how software could genuinely support the way people work together. Harry and I spent many hours in collaborative discussions about workflow: how work actually moves through organizations, where human judgment belongs in a process, and how technology could amplify rather than replace it. More than three decades later, those conversations have scaled up from PC workflow to civilization, and Harry has written the book series those earlier discussions were always pointing toward.

Few people are as well positioned to tackle the topic. Harry earned his PhD with a focus on artificial intelligence at the University of Toronto, one of the birthplaces of the modern field, long before AI was fashionable. He then spent thirty-five years building and leading an enterprise software company serving Fortune 500 customers, while also teaching at the university level and performing as a classically trained singer, including for seniors living with memory loss. That unusual combination, the boardroom, the classroom, and the concert hall, as he puts it, runs through every chapter of this book. It is why his answer to the AI question is never merely technical: he has seen, in three very different worlds, that capability only matters when it is joined to human purpose.

The Human Role in the Age of AI, Vol. 1: The Horizon opens Harry's four-volume series by asking its central question plainly: in a world where machines can do more and more of what once required human intelligence, what is the role of humans? His answer unfolds across a remarkably wide terrain — the three waves of automation and why this one is different; a clear-eyed inventory of AI's capabilities, limits, and risks; time horizons at five, ten, fifteen, and twenty-five years; the economics of productivity, inequality, and a possible AI dividend; and, most distinctively, the questions of meaning: what work is for, what education should form in a person, and what a "meaning society" might look like when scarcity eases. His organizing metaphor is exactly right: the machine may be the engine, but the human must remain the compass.

What sets this book apart from the growing shelf of AI titles is that it is, deliberately, a family book. Harry writes as a researcher, entrepreneur, and teacher, but above all as a father and grandfather, using plain language a grandparent, a mid-career professional, and a student can read together, with practical, generation-specific guidance for each. He is neither a booster nor a doomsayer. He treats AI fear seriously and answers it with responsibility rather than reassurance, and his image of curiosity as the natural voice — powerful in childhood, easily trained into silence, recoverable only with care — is worth the price of the book on its own.

My career has been organized around a single thread, inspired by Doug Engelbart: augmenting human intellect and collaboration with tools, from the workflow systems Harry and I debated in the 1990s through to today's AI assistants. Harry's The Human Role in the Age of AI series arrives as the pace of change has only accelerated, and its guidance is very

Series Preface

The Human Role in the Age of AI

A Four-Book Guide to Helping Every Generation Thrive

Artificial intelligence is not simply another new technology. It is a transformational shift that reshapes work, education, community, and the meaning of our daily lives. Because this transition touches every aspect of our existence, a single conversation—or a single book—is not enough to capture the full scope of our human response.

This four-volume series is designed as an intergenerational roadmap. It explores how we can move through this age of intelligent machines not by surrendering our judgment, but by deepening the very qualities that make us human: curiosity, responsibility, empathy, and wisdom. This first volume provides the essential foundation for the entire journey, mapping the broad terrain we must now navigate together.

- The Human Role in the Age of AI, Vol. 1: The Horizon: Guide for Every Generation Facing the Next 25 Years
- The Human Role in the Age of AI, Vol. 2: The Middle Generation: How Professionals Can Adapt, Lead, and Support Their Families
- The Human Role in the Age of AI, Vol. 3: The Young Generation: How Students Can Learn, Think, and Build a Meaningful Life
- The Human Role in the Age of AI, Vol. 4: The Guiding Generation: How Parents, Grandparents, Teachers, and Mentors Can Help the Young Thrive

The shared question guiding this entire series is simple: As machines become more capable, how can human beings become wiser, deeper, and more responsible in response? We are not merely surviving a technological shift; we are building a civilization that must remain answerable to human dignity and purpose. I hope this series serves as your compass as we navigate this journey together.

Preface

The Role of Humans in the Age of Artificial Intelligence

A Very Unusual Moment in History

We are living in a very unusual time in human history. Throughout the past, when new technologies were invented, they primarily replaced human muscle. Machines became stronger than we are. Tractors replaced farm labor, industrial machines replaced factory labor, and computers replaced portions of office work. Each of these transformations changed how people lived and worked, but they all shared one underlying characteristic: they extended human physical capability or improved efficiency.

What is happening today is different in an important way. For the first time, we are building machines that can perform tasks that once required human intelligence. They can write, translate, analyze, generate images, draft code, compare documents, and increasingly assist in decisions that used to belong only to educated human labor.

This leads us to a new kind of question — one that previous generations did not have to face in quite this form: if machines can do more and more of what once required thinking, what is the role of humans? That is the central question of this book. It is not mainly a technical question. It is a human question.

My Perspective on This Shift

My perspective on this shift comes not only from observing it, but from living through multiple phases of technological change. I began my career in artificial intelligence at a time when the field was far from mainstream attention. Later, I worked as a teacher and as an entrepreneur, and over the years I have spoken with many different communities about what technological change means in real human terms.

When I speak about AI to community groups, especially older adults and parents, I often sense the same mixture in the room: curiosity, anxiety, and a desire for plain language. People are not asking for a technical lecture on model architecture. They want to know how to protect their families, how to remain useful, and how to help young people face a future that feels less predictable than before.

In my own work in AI, teaching, and business, I have seen the same pattern again and again: technology changes quickly, but human questions do not. People still ask whether they matter, whether their children will be prepared, whether society will remain humane, and what kind of life is still worth building when the world changes faster than expected.

That is one reason I wrote this book. I did not want to write only about systems, models, and tools. I wanted to write about the human meaning of the transition.

Four kinds of personal experience shape that human meaning for me. I have watched young children, including my own young grandchildren, approach the world with natural curiosity and joy. I have taught college and graduate students and seen how much they need breadth, clear writing, clear speaking, and awareness of the wider world. I have built a business, worked with employees, and dealt with major corporate customers where technical skill mattered, but trust, communication, judgment, and human relationship often determined how far people could go. And I have sung for older adults, including people with memory loss, and seen how a familiar song can awaken dignity and connection in ways no machine can fully replace.

Understanding Fear Through a Familiar Lens

Before we go further, we must address a deeper issue: the fear and uncertainty many people feel about artificial intelligence. After many of my talks, people often come up to me with questions that are not technical. A professional may ask, "Will my experience still matter?" A parent may ask, "What should my child study now?" A senior may simply say, "I don't understand this, and it worries me."

These are not questions about algorithms. They are questions about stability, identity, and the future. This reaction is natural. Whenever a new technology appears that seems powerful and difficult to control, it challenges our sense of stability.

To better understand this fear, it is helpful to look at a familiar example: the automobile. Imagine if the car were invented today. We would see a powerful machine capable of traveling at speeds far beyond human limits. We would recognize that, in the wrong hands, it could be dangerous — even deadly. If we focused only on these risks, we might conclude that such a technology should be restricted.

But that is not what happened. Society made a different choice. We recognized the immense value of the automobile and chose to manage its risks rather than eliminate it. We built an entire ecosystem around it — traffic laws, safety systems, licensing rules, road design, and cultural norms.

Artificial intelligence should be understood in a similar way. It is not just another tool. It is, in an important sense, an engine of the mind. Like cars, AI carries risks. But its potential benefits may be far greater. The goal is not to stop the technology. The goal is to ensure that we remain in the driver's seat.

The Three Waves of Automation

Wave 1 - Mechanize Muscle: Machines amplified strength and replaced much physical effort.

Wave 2 - Digitize Information: Computers automated calculation, records, communication, and processing.

Wave 3 - Automate Cognition: AI reaches into drafting, explanation, analysis, judgment support, and cognitive work.

In this new wave, the focus moves from performing tasks to making judgments, from executing processes to defining direction. As machine capability rises, human value moves toward interpretation, responsibility, communication, and human presence.

Beyond Technology: A Human Question

This transformation is not only about technology. It is also an economic question, an educational question, a social question, and ultimately a human question. While there are many books that explain how artificial intelligence works, this book is about something deeper.

It is about our relationship with this technology. It is about how we maintain agency in a world where machines can increasingly act on our behalf. It is about how we preserve dignity in a world where productivity may no longer be tied to labor in the old way. And it is about how we educate human beings to think with clarity and live wisely when powerful systems can increasingly supply answers for them.

A Guiding Theme

There is one central idea that will appear throughout this book: the future of AI is not only about what machines can do, but about what humans should do. As machines take on more tasks, we must think more carefully about the purpose of education, the meaning of work, the distribution of dignity, the structure of opportunity, and the qualities that truly matter in human life.

These are not technical questions. They are human questions. And in many ways, they are the questions that determine whether a civilization becomes wiser as it becomes more powerful — or merely more capable without becoming more humane.

Who This Book Is For

This book is written for individuals concerned about their future, for parents thinking about their children, for students planning their education, and for educators reconsidering how to teach. It is also written for older adults who do not want to be pushed to the margins of a technological age, and for middle-aged adults trying to understand what this transformation means for work, family, and practical life.

You do not need a technical background to read this book. It is written in clear language and focuses on ideas rather than technical complexity. The goal is not to turn the reader into an engineer. The goal is to help the reader think with greater clarity about what kind of life, work, education, and society the age of AI may now require.

The Question Behind the Journey

For me, this question is not theoretical. I have lived long enough to see several technological waves arrive with great promises, great fears, and great unintended consequences. In every wave, the tool mattered, but the human response mattered more. The same will be true of AI. The future will not be decided only inside laboratories or boardrooms. It will also be decided at kitchen tables, in classrooms, in workplaces, in senior communities, and in the quiet choices families make about what to value.

That is why I have written this volume in plain human language. I want a grandparent, a professional, a student, and a teacher to be able to sit with the same book and recognize their own concerns inside it. The AI age is large, but it enters life through ordinary doors: a child choosing a major, a worker trying to adapt, an older adult trying not to be left behind, and a family asking what still makes a life meaningful.

Every chapter in this book is an attempt to answer a single question: in a world where machines can do more and more, what is the role of humans?

Because in the end, the most important question may not be how intelligent machines become, but how humans choose to live — and what kind of civilization they choose to build — in response.

Part I Introduction

Understanding the AI Revolution and Its Physical Horizon

This opening part gives the reader orientation: what AI already does, where it may be heading, and why its risks require human responsibility rather than panic.

Chapter 1

The Arrival of Artificial Intelligence

Horizon Insight: AI is not merely a new product category. It changes the conditions of ordinary life, which means the human response must begin with clarity rather than panic.

A Rare Moment in Human History

Many people first encounter artificial intelligence not as an abstract theory, but as a small surprise. A grandparent sees a computer explain a medical term in plain language. A student asks for help understanding a difficult chapter. A professional watches a system draft in seconds what once took an hour. At first, the reaction is often simple curiosity. Then a deeper feeling appears: something important has changed.

That feeling is worth taking seriously. This chapter is not meant to create fear, and not to exaggerate what AI can already do. It is to help the reader pause long enough to see that this is one of those moments when a tool begins to change the conditions of ordinary life.

Every few hundred years, humanity experiences a technological change large enough to reshape society at its foundations. The agricultural revolution transformed human life from nomadic survival into settled communities. It allowed people to store food, build villages, organize labor, and create the first enduring civilizations. The industrial revolution transformed work by amplifying human strength, reorganizing production through factories, and drawing populations into cities. The computer revolution transformed information. It changed how we calculate, communicate, store knowledge, and connect across the world.

Today, we may be standing at the beginning of another major shift: the age of artificial intelligence. The change may reach deeply into ordinary life. Previous revolutions changed where people lived, how people worked, and how information moved. Artificial intelligence may change not only what we do, but how we think, how we learn, how we organize society, and even how we understand the distinctive role of human beings.

That is why this moment deserves to be taken seriously. It is not simply the arrival of a new product category or a faster piece of software. It may become one of the major shifts in how human beings learn, work, communicate, and organize daily life.

Why Families Feel Anxious

Lately, many dinner-table conversations have become unusually anxious. Parents look at children studying computer science, journalism, business, design, or medicine and wonder whether the path they once trusted is still safe. Should they push a child to change majors? Should they question the value of college? Should they steer a gifted young person toward a trade simply because the headlines make every digital profession sound fragile?

That anxiety does not mean parents are paranoid. It means they care deeply. The traditional promise—good grades, a good college, a respectable first job, and a stable career—has been weakened by a world in which machines can now draft, summarize, code, analyze, and produce polished work in seconds.

But fear is a poor architect. Decisions made in panic often trade long-term human potential for a short-term illusion of safety. Before a family uproots a young person's education or career direction out of fear, we have to ask a deeper question: not which job title looks safest today, but what kind of human being will remain adaptable, trusted, and useful when the job market of 2035 looks different from anything we can predict now.

Our job as parents, grandparents, teachers, and mentors is not to predict the unpredictable job market of 2035. Our job is to help cultivate adaptable, deep, and resilient human beings who can navigate whatever that market becomes.

From Science Fiction to Daily Reality

For many older readers, AI may still carry the flavor of science fiction. For younger readers, it may already feel like part of everyday life. That difference between generations is itself important. One generation is watching an imagined future arrive; another is growing up inside it.

For many years, artificial intelligence belonged mainly to the world of imagination. People encountered it through science fiction: intelligent robots, talking machines, systems that could reason like humans, and computers that might one day become companions or rivals. These stories were powerful because they touched something deep in the human mind. They raised questions about consciousness, control, identity, and the boundary between tool and being.

But for most people, these ideas still felt distant from everyday life.

In practice, early artificial intelligence was narrow and limited. Computers could calculate very quickly and follow instructions with precision, but they could not reliably understand natural language, recognize images the way humans do, or carry on meaningful conversations. They were impressive machines, but they were still clearly machines. They did exactly what they were told, and no more.

That assumption is now being challenged. The dream of artificial intelligence is old, but the experience is newly widespread. Millions of ordinary people can now use capabilities that once seemed far away. They can ask a system a question in natural language and receive a coherent response. They can upload an image and get an explanation. They can request a draft, summary, lesson, translation, comparison, or piece of code and receive it immediately. Science fiction has not fully arrived, but it has stopped feeling entirely fictional.

Why This Moment Feels Sudden

In recent years, artificial intelligence has advanced in a way that feels visible and dramatic. Today, AI can recognize objects in images, understand and generate human language, translate between languages, write essays and reports, generate images and videos, help write computer programs, assist in diagnosis, and increasingly act as both tutor and assistant.

For many people, this progress feels sudden, almost as if it appeared overnight. In reality, it did not. Artificial intelligence has been developing for decades through work in computer science, statistics, mathematics, engineering, neuroscience-inspired ideas, and large-scale digital infrastructure. What we are seeing now is the moment when long-term progress becomes visible all at once.

When a technology remains inside laboratories, research papers, and specialized industries, most people do not feel its importance. But when it begins to enter daily life — when students use it, professionals use it, businesses use it, and families begin talking about it — the change suddenly feels immediate. That is why the current moment can feel both exciting and unsettling. Something that seemed abstract has become practical. Something that seemed distant has become personal.

Why Now: A Convergence of Forces

Several different forces matured at the same time and created a tipping point. Advances in computing power made it possible to process enormous amounts of information more quickly than before. The availability of vast amounts of digital data allowed systems to learn from real-world examples on a massive scale. Improvements in algorithms made learning more effective and flexible. Cloud computing made these capabilities widely accessible rather than locked inside a few specialized institutions. Global connectivity allowed models, ideas, and applications to spread at extraordinary speed.

Each of these developments matters on its own, but their true importance lies in their convergence. A single improvement rarely changes the world. A cluster of improvements that reinforce one another can. Better hardware supports larger models. More data supports better learning. Better algorithms make data and hardware more useful. Cloud systems make the resulting tools accessible to millions. Once that cycle begins, capability stops being isolated and starts becoming part of ordinary infrastructure.

From Tools to Assistants

One of the clearest ways to understand the current shift is to look at the changing relationship between humans and computers. In the past, computers were tools that required precise, step-by-step instructions. If you wanted a task completed, you had to specify exactly what to do and often exactly how to do it. The human being had to translate intention into explicit procedural language.

Today, that relationship is beginning to change. Instead of always telling the computer how to do something, you can increasingly tell it what you want and let the system attempt to determine the steps for you. You can ask it to summarize a long report, explain a difficult concept, create a presentation outline, compare two plans, help draft a letter, organize research, or support a complicated decision.

This is not merely a matter of convenience. It represents a profound shift in the history of computing. The user increasingly works at the level of goals, while the machine increasingly works at the level of execution. That changing surface relationship will have major implications for productivity, education, and work.

The main lesson is that collaboration with AI begins with human purpose, not with blind automation.

A Subtle but Powerful Transformation

For decades, humans had to adapt themselves to computers. They had to learn rigid interfaces, memorize commands, understand software logic, and work within narrow systems designed around machine requirements. In that older model, the burden of translation fell mostly on the human.

Now the burden is beginning to shift. Computers are increasingly adapting to humans by understanding natural language, broad intention, and patterns of use. This lowers barriers to entry. It allows people who are not programmers or technical specialists to access capabilities that would once have been inaccessible to them.

That changes what counts as valuable human effort. If the machine can increasingly help with execution, then the human contribution moves upward toward asking the right question, defining the real goal, judging what is accurate, deciding what is worthwhile, and interpreting consequences. This is one reason AI feels different from many earlier tools. It is not only helping us do old tasks faster. It is changing the division of labor between person and machine.

Why This Time May Be Different

In previous technological revolutions, machines replaced or amplified physical labor. Later, computers replaced calculation and supported information management. Automation reduced repetitive tasks. Artificial intelligence goes further because it reaches into cognitive work itself. It can now participate in activities long assumed to belong mainly to educated human beings: drafting, summarizing, explaining, translating, comparing, and generating.

That is why this moment may be different from what came before. The affected zone is no longer limited to factories, warehouses, or routine clerical work. It now extends into the professions, education, creative work, administration, law, medicine, design, software, consulting, research, and communication.

This does not mean the machine simply replaces the human being in a clean one-to-one way. Real life is always more complicated than that. But it does mean that the old comfort zone is weakening. The assumption that cognitive work is automatically safe simply because it is intellectual is no longer reliable.

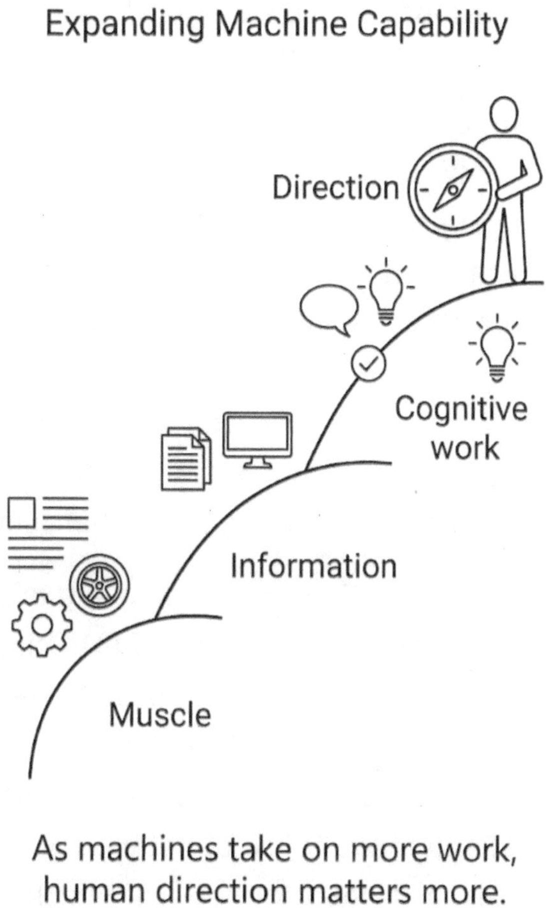

Expanded machine capability increases, rather than removes, the need for human judgment.

The Entry-Level Catch-22

Many professions have traditionally trained beginners through junior tasks: first drafts, basic research, routine coding, document review, summaries, scheduling, and standard analysis. If employers automate too many of those beginner tasks, young people may be told they need experience before being hired, while losing the very entry-level work through which experience used to be gained. This is the Entry-Level Catch-22.

The solution is not to preserve meaningless busywork. It is to build new bridges into real experience: supervised projects, apprenticeships, internships, proof-of-work portfolios, and junior roles where beginners learn to direct AI responsibly while still developing judgment, communication, and domain understanding. That training challenge is only one consequence of rapid progress. A second consequence is that AI can increasingly contribute to the development of the very systems that come after it.

A Technology That Can Help Improve Itself

Another important characteristic of artificial intelligence is that it can contribute to its own improvement. AI systems can help design better computer chips, write more efficient software, analyze scientific research, accelerate engineering workflows, and identify useful patterns across many fields. This creates a feedback loop in which better AI can contribute to the development of still better AI.

That possibility matters because it suggests that progress may not remain slow and linear. Instead, we may experience periods of acceleration. Each new improvement can strengthen the environment in which the next improvement is made. When that happens, society's challenge is no longer only to invent. It is also to adapt quickly enough to the pace of invention.

This is one reason many people feel unsettled. Human beings and institutions adapt more slowly than digital systems. Schools, laws, careers, cultural expectations, and moral frameworks do not update at software speed. So, the issue is not merely that AI is improving. It is that the rest of society may struggle to keep pace with its improvement.

Not Just a Technology Story

We can easily mistake artificial intelligence as a purely technical development. In reality, its significance extends far beyond technology. It affects how people work, how they learn, how wealth is created and distributed, how companies compete, how nations position themselves, how social systems evolve, and how individuals understand meaning and identity.

In this sense, artificial intelligence is not just a technology story. It is also an economic story, an education story, a work story, a family story, and ultimately a human story. Whenever a major technology arrives, society eventually asks deeper questions. What becomes more valuable? What becomes less valuable? Who gains power? Who loses security? What skills matter now? What happens to dignity and purpose when tasks shift from human beings to machines?

Those are not engineering questions alone. They are civilizational questions.

We should remember that we are at the beginning of this transformation, not the end. The systems visible today are early versions of a longer process that will unfold over years and decades. The tools will improve. Their costs will fall. Their reach will widen. Their effects will accumulate. Some people and institutions will use the shift well; others, especially those built mainly on routine execution, will feel the pressure sharply.

Panic is not useful, but complacency is not wise either. The right response begins with clarity. We need to see the scale of the shift, understand why it is happening now, and begin asking not only what machines can do, but what kind of human future we want to build in their presence.

Chapter 2

What Artificial Intelligence Can Do Today

Horizon Insight: AI is powerful because it can work across language, images, sound, code, analysis, learning, and action. But fluency is not wisdom; the human being must still supply purpose, verification, and judgment.

From Abstraction to Reality

The easiest way to understand AI today is to begin with ordinary scenes. A retiree asks a chatbot to explain a confusing insurance letter. A teacher asks for examples to help students understand a difficult concept. A small business owner asks for a first draft of an announcement. A student asks for help practicing a weak subject. None of these moments looks dramatic from the outside, yet together they show how quickly AI is moving from abstraction into daily life.

In my own lectures, I have seen this shift happen in real time. A room may begin with polite curiosity, but when people see AI summarize, explain, translate, draft, or compare ideas in front of them, the conversation changes. It is no longer a faraway technology story. It becomes a practical question about daily life, work, learning, and trust.

To understand the social impact of AI, we first need a grounded sense of what it can already do. AI is not a conscious mind, but it is far more than a clever spellchecker or search box. It is a family of systems that can assist with language, images, sound, code, analysis, learning, and increasingly multi-step action. Used well, it can help people begin. Used carelessly, it can make people trust fluent output too quickly. That distinction will matter throughout the rest of this book.

A Plain-English Note on How AI Actually Works

Before looking at the different things AI can do, it helps to pause for a plain-English explanation of what is happening underneath. Readers do not need to become engineers to understand the basic idea. But without at least a rough picture, AI can seem either magical or frightening. Neither reaction helps us think with clarity.

Most modern AI systems learn from patterns. They are trained on enormous amounts of text, images, code, sound, and other examples. From that training, they learn statistical relationships: which words often follow other words, which visual shapes tend to belong together, which code structures solve certain kinds of problems, and which patterns usually appear in a given context. When we ask a question, the system does not search its memory the way a person recalls a childhood story. It generates a response by estimating, step by step, what output is most likely to fit the prompt and the patterns it has learned.

For that reason, AI can look so impressive. Human language itself contains a great deal of structure. If a system has absorbed enough patterns from books, articles, websites, conversations, software, and other forms of human expression, it can produce answers that feel fluent, organized, and sometimes surprisingly insightful. It can imitate the shape of explanation, argument, design, code, and even certain kinds of creativity because it has learned many patterns in how human beings express knowledge.

But this is also why AI must be used carefully. Pattern prediction is not the same as lived understanding. A system can describe grief without having grieved. It can write about music without feeling breath move through the body before a difficult phrase. It can discuss ethics without carrying conscience. It can produce a medical-sounding answer without being a doctor who must face the patient. It can sound certain without actually knowing in the human sense.

Another way to say it is this: AI is powerful at producing plausible output, but plausibility is not the same as truth. Sometimes the output is excellent. Sometimes it is incomplete. Sometimes it is wrong in subtle ways. Sometimes it repeats patterns from its training data that should be questioned rather than accepted. And because the writing can sound smooth, the danger is that people may trust it too quickly.

This does not make AI useless. It makes human judgment more important. If we understand that AI works by learning and generating patterns, we can use it with the right attitude. We can ask better questions, compare answers, request sources, check important claims, and decide when a human expert is needed. We can treat AI as a powerful assistant, tutor, drafting partner, and pattern finder, but not as a final authority or moral actor.

This simple understanding will help throughout the rest of the chapter. AI can work with language, images, voice, video, code, analysis, agents, and education because these areas all contain patterns that machines can learn and generate. But in every one of these areas, the human being remains responsible for purpose, verification, judgment, and meaning.

AI Can Understand and Generate Language

One of the most important breakthroughs of recent years is the ability of AI to work with human language. Language sits near the center of modern life. It is how we communicate ideas, learn new concepts, coordinate work, explain decisions, persuade others, and express imagination. A large portion of knowledge work consists not of physical labor, but of reading, writing, summarizing, interpreting, drafting, and responding.

Today, AI systems can answer questions, explain complex topics, write essays and reports, summarize long documents, translate between languages, assist in writing computer code, and help generate ideas. What makes this capability so striking is not only the quality of many outputs, but also the speed and accessibility. Tasks that once required hours of concentrated work can now be started in seconds and often completed with assistance in minutes.

A business professional can draft a report far more quickly than before. A student can ask for an explanation matched to a particular level of understanding. A writer can explore multiple angles on a problem without beginning from a blank page each time. A researcher can compare summaries of complex material before deciding where to look more deeply. In each case, AI does not remove the need for human judgment, but it changes the starting point of the work. Increasingly, people begin with an AI-generated response or draft and then refine, verify, challenge, or improve it.

A Shift in Cognitive Work

That is why the change can feel both helpful and unsettling. The same tool that saves time also asks us to rethink what human skill now means. If the first draft is easy, then the real human work becomes clearer: judgment, accuracy, taste, responsibility, and the ability to know what should be said in the first place.

This change is not merely about convenience. It represents a shift in how cognitive work is organized. In the past, language-based work required time, experience, and repeated exposure. People developed expertise slowly through practice. They learned how to write well, summarize accurately, structure arguments, and explain difficult ideas because they had to do those things themselves over and over.

This shift in cognitive work becomes clearer through another familiar image: driving. A large language model is not a conscious brain; it is more like a powerful cruise-control system for language. If the driver gives no destination and pays no attention, the car simply follows the most ordinary road ahead. It may move smoothly, but it does not know where life ought to go.

The human being must remain the driver. We provide the destination, the judgment, the ethics, and the sense of purpose. AI can help keep pace, suggest turns, and reduce some strain, but it must not decide the journey for us.

Now, artificial intelligence lowers the barrier to entry for many of these tasks. More people can produce a draft. More people can generate an explanation. More people can explore a topic quickly. This can be empowering, but it also raises a more serious question. If AI can generate language and assist with thinking, then what becomes the distinctive role of human intelligence?

One early answer is that the human role shifts upward. If AI increasingly helps with drafting and first-pass organization, then the human being becomes more responsible for asking the right question, defining the real goal, judging what is accurate, deciding what is worthwhile, and shaping the final meaning.

AI Can Recognize and Generate Images

Artificial intelligence is no longer limited to text. Modern systems can process and generate visual information as well. They can recognize objects and faces in images, read handwritten text, analyze certain kinds of medical images, identify patterns in photographs, and generate realistic or stylized visuals from a simple written description.

This has real consequences across many fields. In healthcare, AI can help doctors detect patterns in scans that may not be visible at first glance. In design, architecture, and marketing, it can quickly generate multiple visual concepts based on a brief description. In education, it can help create explanatory visuals. In creative industries, it allows individuals to explore different directions far more quickly than before.

What changes here is not only productivity, but the nature of the creative process itself. Instead of manually creating every version of a concept from scratch, a person can generate many possibilities, compare them, reject weak directions, and refine stronger ones. In many cases, the human being spends less time producing the first rough version and more time selecting, interpreting, judging, and improving.

AI Can Generate Voice and Video

The expansion of AI beyond text and images into audio and video makes the picture even more striking. Today, systems can generate realistic human voices, translate speech into other languages, create digital avatars, transcribe conversations, assist with dubbing, and help with video editing and content creation. Capabilities that once required specialized studios, significant budgets, or teams of experts are becoming much more accessible.

Taken together, these capabilities open new possibilities in communication, education, business, and media. Educational material can be adapted for different languages and audiences much more quickly. Businesses can create personalized communication at scale. Content creators can produce higher-quality materials with fewer resources. Organizations can experiment more rapidly with explanation, training, and outreach.

At the same time, these capabilities create serious new challenges. As generated voices and videos become more convincing, it becomes harder to distinguish real content from fabricated content. Impersonation becomes easier.

Scams become more sophisticated. Trust becomes more fragile. This means society will need not only technical tools, but also new habits, standards, and norms to preserve accountability.

Because AI now crosses text, image, voice, and video, a single diagram can help the reader see the breadth of the change.

Across text, images, voice, and video, AI may generate in many media, but human beings must still decide what is true, appropriate, beautiful, and trustworthy.

AI Can Help Write Computer Programs

One of the most economically significant capabilities of AI today is its ability to assist in software development. AI can suggest code, identify bugs, explain unfamiliar functions, generate templates, refactor sections of a program, and help non-specialists build basic applications or scripts that would once have required far more expertise.

For programmers, this can greatly accelerate workflow. For non-programmers, it lowers the barrier to experimentation. The effect is not only that software gets written faster. It is also that more people can participate in software-like thinking. That matters because software increasingly shapes every sector of life, from healthcare and logistics to education and finance.

This does not eliminate the need for skilled engineers. In many cases, it increases the value of deep engineering judgment because generated code still needs to be checked, integrated, tested, and understood. But it does change how coding work begins and how much leverage one person can have.

AI Can Support Analysis and Decision-Making

AI is also becoming useful in tasks that involve comparison, pattern recognition, and structured analysis. It can sort through large sets of documents, summarize trends, identify anomalies, highlight key differences, and present organized options for review. This is particularly useful in law, business, medicine, administration, policy work, and research.

Again, the important point is not that AI makes the final decision in the human sense. The important point is that it changes what the human decision-maker does. Instead of spending most of the time assembling first-pass information, the person may spend more time judging, questioning, prioritizing, and deciding what to do with the information once it is organized.

That shift changes the center of many professional roles. As more preparation work becomes machine-assisted, the human value moves toward judgment and accountability.

AI Can Act as an Agent

Another important capability is now emerging: AI can increasingly act as an agent. A chatbot mainly answers a question. An assistant helps with a task. An agent goes one step further. It can be given a goal, break that goal into steps, use tools, interact with software or websites, and report progress back to the human user.

This distinction matters because it changes the practical meaning of artificial intelligence. A person may not simply ask AI to draft a message. They may ask it to research options, compare prices, prepare a schedule, organize documents, summarize responses, fill out a form, draft follow-up emails, or coordinate a sequence of work across several tools. In business settings, an agent may help monitor a workflow, prepare routine reports, organize customer information, or move a process from one step to the next under human-defined rules.

The important point is that AI is beginning to move from answer generation toward goal-directed action. That does not make it independent in the human sense. A responsible agent still needs boundaries, permissions, verification, and human accountability. But it does mean that the technology is becoming more operational. It can increasingly participate not only in thinking about work, but in carrying out parts of the work.

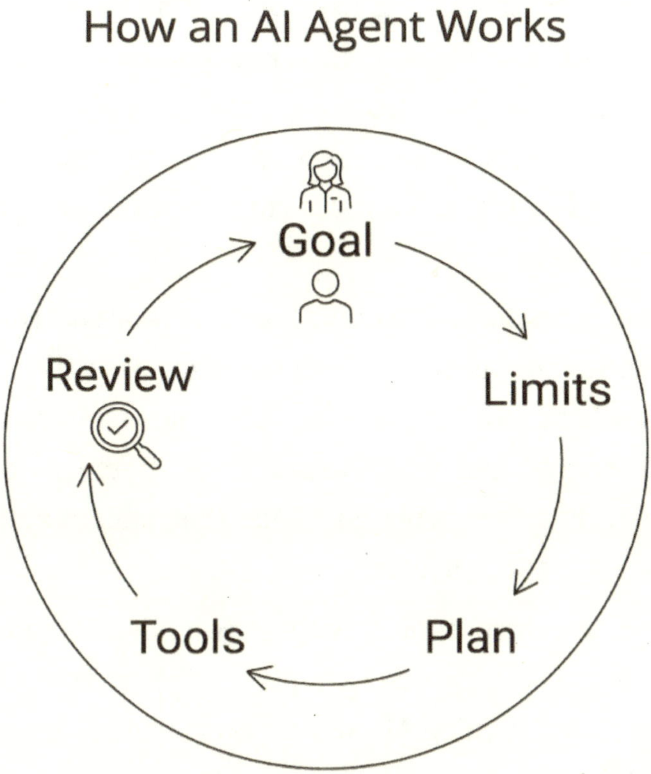

An agent may act in steps, but people set limits and review results.

A step-by-step AI helper still needs human goals, boundaries, and review.

AI Can Be a Tutor and a Companion for Learning

One of the most promising uses of AI today is educational support. An AI tutor can explain a concept at different levels, answer follow-up questions, create practice exercises, generate examples, translate explanations into another language, and continue a patient conversation for as long as the learner needs.

This has the potential to make learning more personalized than traditional classroom systems often allow. A student who is confused can keep asking. A learner embarrassed to speak in class can still explore. A person studying independently can receive guidance that once would have been much harder to access.

Yet this promise comes with a warning. A tutor that always provides answers can also weaken the habit of wrestling with difficulty. AI can deepen learning, but only if it is used to extend thought rather than replace it. The educational value of AI therefore depends not only on the technology itself, but on the philosophy with which it is used.

What AI Still Does Poorly

To understand what AI can do today, it is equally important to understand what it still does poorly. AI can sound confident when it is wrong. It can miss context that matters deeply. It can generate plausible nonsense. It can reflect biases in the material from which it learned. It often lacks the stable commonsense grounding that human beings develop through lived experience in the world.

It also does not carry responsibility in the human sense. A machine can produce a recommendation, but it does not bear the moral burden of the consequences. That remains with the human being. That is why the growth of AI does not remove the need for human oversight. In many cases, it increases the need for it.

So, we need a balanced view. AI is already powerful enough to reshape workflows and expectations. But it is not magical. It is not infallible. It is not a substitute for conscience, judgment, or accountability.

Why These Capabilities Matter So Much

Taken together, these capabilities explain why the present moment feels different from earlier technology moments. AI is not improving only one narrow function. It is operating across language, image, sound, code, analysis, learning, and increasingly agent-like action. It touches multiple forms of human expression and multiple layers of cognitive work at once.

That breadth is one reason the technology feels so consequential. A single system can now assist with communication, design, programming, planning, research, education, and the coordination of multi-step tasks. That multiplies its effect. It is not just a specialized tool for one profession. It is becoming a general-purpose cognitive and operational partner across many domains.

This does not mean the future is predetermined. But it does mean the scale of the change is already large enough that we should stop treating AI as a distant possibility. It is already here, already useful, already imperfect, and already changing the starting point of human work.

Taken together, if Chapter 1 established that artificial intelligence is a rare historical development, this chapter has shown why that claim should be taken seriously. AI can already work with language, image, sound, code, analysis, learning, and increasingly agent-like workflows in ways that would have seemed extraordinary not long ago. These are not isolated tricks. They are the beginnings of a broader transformation in how cognitive and administrative work is performed.

Chapter 3

What AI May Be Able to Do in 5, 10, 15, and 25 Years

Horizon Insight: Horizon thinking reduces panic. We cannot predict the future precisely, but we can prepare more wisely when we see the likely direction of change.

Why Time Horizons Matter

Reading the Future as a Family

A horizon is most useful when it becomes a family conversation rather than an abstract forecast. Around a dinner table, the five-year question may be: what should my child learn now? The ten-year question may be: how will my profession change? The fifteen-year question may be: how can my parents remain independent? The twenty-five-year question may be: what kind of society do we want our grandchildren to inherit? When a forecast is translated into those human questions, it stops being speculation and becomes preparation.

That is why I do not treat horizon thinking as prediction in the narrow sense. I treat it as a discipline of attention. It helps us see that the future is not one event. It is a sequence of pressures arriving at different speeds. Families, schools, companies, and communities do not need perfect certainty. They need enough orientation to stop reacting only to headlines and start building habits, skills, and institutions before the pressure becomes overwhelming.

Looking into the future should not be an exercise in panic. It should be an exercise in preparation. When I speak with audiences about AI, I often see that people are less frightened when they can place change on a timeline. A vague future feels overwhelming. A set of horizons gives the mind something to hold.

The horizons in this chapter should therefore be read as guideposts, not prophecies. They are meant to help families, workers, educators, and older adults prepare with steadiness rather than react with confusion.

The goal is orientation rather than prophecy. If we understand the likely direction of change, we can prepare more thoughtfully and avoid reacting only after disruption has already arrived.

That is why it is useful to think in time horizons. A five-year horizon helps us think about what is likely to become normal in the near term. A ten-year horizon helps us imagine what happens when current tools become more agent-like and more deeply embedded in institutions. A fifteen-year horizon helps us think about AI moving further into health, aging, logistics, caregiving, and the physical world. A twenty-five-year horizon pushes us toward much larger questions about work, identity, social order, meaning, power, and the human future itself.

None of these horizons is exact. But together they create a map. And when change is large enough, having a map matters even if the exact road still contains uncertainty.

A singing metaphor may help. In classical music there is a beautiful idea called tempo rubato, or stolen time. A singer does not perform a line like a mechanical metronome. She may hold back a fraction of a beat before an important word, then move forward again so the phrase still belongs to the larger structure of the song. The tempo bends, but the music does not collapse.

That is also how I think society must approach the next twenty-five years of AI. We cannot live by a rigid script, because the global tempo will change. Some shifts will arrive sooner than expected; others will take longer. Families, schools, businesses, and governments need enough structure to stay grounded, but enough rubato to adjust the rhythm when the world moves unexpectedly.

Thinking About the Future Without Pretending to Be Certain

It is difficult to predict the future of technology because technological progress is not driven by one variable alone. It depends on scientific breakthroughs, engineering progress, business incentives, regulation, infrastructure, energy, politics, adoption behavior, and many other forces. Even when a technology is possible, that does not mean it will spread quickly. And even when it spreads quickly, that does not mean society will adapt at the same speed.

For that reason, it is wiser to think in terms of possibility, pressure, and direction rather than certainty. Instead of asking what definitely will happen, we ask what may happen if current trajectories continue and if key enabling factors mature. This does not eliminate uncertainty, but it creates a more useful mindset. It shifts us from false confidence to strategic preparedness.

People often make two opposite mistakes. One group assumes that because precise prediction is impossible, forward thinking is pointless. Another becomes intoxicated by certainty and treats imagined futures as if they were guaranteed. Both mistakes are unhelpful. The better path is disciplined imagination grounded in current capability and historical pattern.

The Next 5 Years: The Work Shock Becomes Visible

In the next five years, artificial intelligence will likely become a normal part of everyday institutional life. We are already entering what might be called the integration phase. Instead of existing as a separate novelty, AI will increasingly be embedded inside email systems, office software, search platforms, productivity suites, design tools, customer service flows, programming environments, educational platforms, and healthcare workflows.

During this period, AI will act primarily as a powerful assistant. It will help people draft documents, generate presentations, analyze data, summarize large amounts of information, plan activities, compare options, produce content, and manage routine mental work much more quickly. Tasks that once required significant time may increasingly take only minutes. Individuals will appear more productive, not because they suddenly became more knowledgeable overnight, but because they have access to a cognitive amplifier.

But the real issue in the five-year horizon is not only convenience. It is the changing structure of work. Many tasks currently performed by junior knowledge workers are already exposed: first-draft writing, document comparison, standard reporting, basic legal review, routine coding, scheduling, customer support, research summaries, and many forms of administrative coordination. As these functions become cheaper and faster, the old entry path into many professions may begin to narrow.

This is the Entry-Level Catch-22 defined in Chapter 1. In the five-year horizon, it becomes visible as AI compresses junior knowledge-work tasks and forces schools, employers, and families to build new bridges into real experience.

The five-year horizon may also flatten some layers of middle management. If senior people can use AI to interact with more information, supervise more processes, and coordinate more workers directly, some intermediate organizational layers may weaken. This does not mean all management disappears. It means management itself changes. More roles may be judged by judgment, communication, and change leadership rather than routine information handling.

So, the first horizon is not mainly about machine replacement in the dramatic science-fiction sense. It is about compression. Compression of time. Compression of routine tasks. Compression of beginner roles. Compression of some career ladders. That is enough to create major social and psychological consequences even before the deeper future arrives.

The 10-Year Horizon: From Assistants to Agents

Within ten years, the more important shift may be from assistants to agents. An assistant responds to instructions one step at a time. An agent can be given a broader goal and can determine a sequence of actions needed to pursue that goal. This may sound like a technical distinction, but it has major consequences for work, home life, education, and organizations.

A simple way to understand the shift is this: a chatbot answers, an assistant helps, and an agent acts within boundaries. Instead of asking AI to perform one isolated task after another, people may increasingly assign broader objectives such as organizing a project, maintaining a workflow, managing part of a hiring process, coordinating a campaign, preparing a travel plan, or handling routine administrative steps within specified limits.

The system may plan steps, use software tools, search information, prepare drafts, monitor progress, send reminders, compare options, and make bounded adjustments while the human supervises at a higher level. In business, this may turn AI from a drafting tool into a workflow participant. In daily life, it may turn AI from a question-answering system into something closer to a personal administrative helper.

This changes management itself. A growing number of people may find themselves not only managing humans, but also managing systems of AI agents. The practical question becomes less, “How do I execute this task?” and more, “How do I define the right goal, set the right boundaries, supervise the system intelligently, and judge whether the result is acceptable?”

The ten-year horizon therefore deserves to stand on its own rather than being treated merely as a bridge paragraph between the near term and the longer term. It marks a real shift in the structure of work: from machine assistance inside human workflows toward bounded machine autonomy inside human-defined systems.

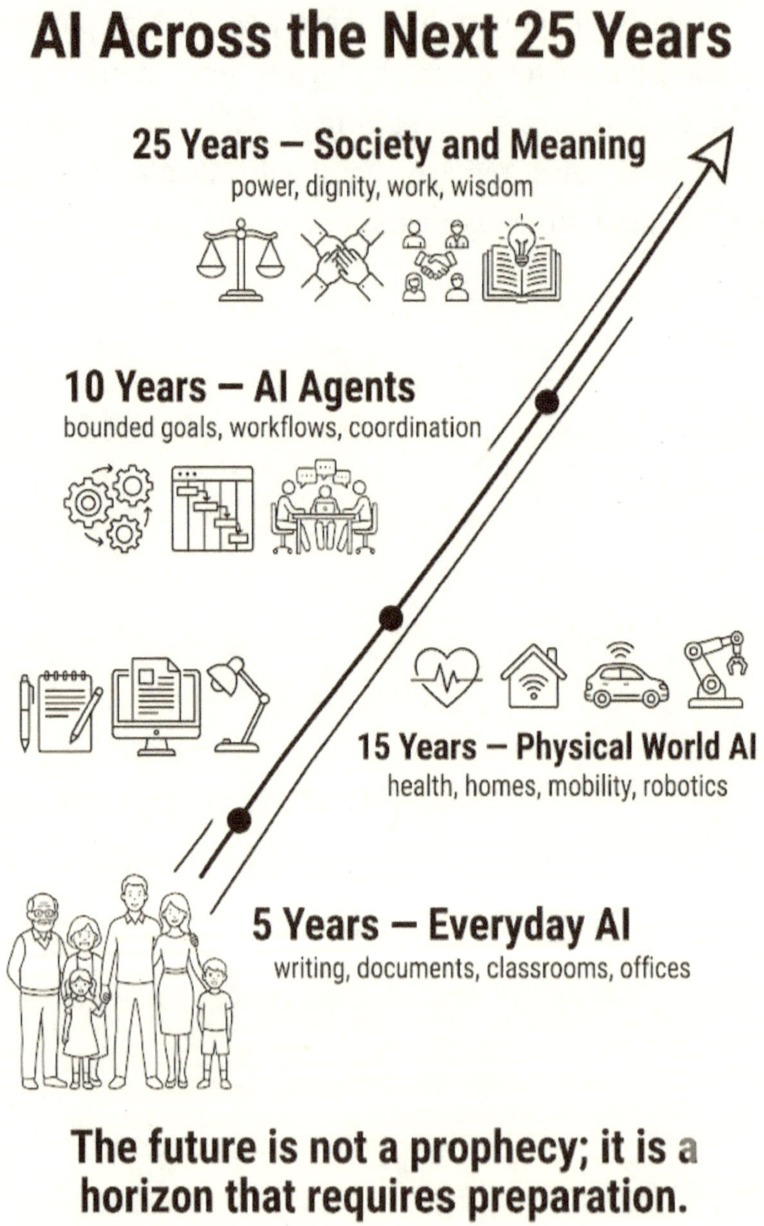

Goal-driven AI requires a different kind of supervision: the tool may carry out steps, but the human still defines the purpose and checks the result.

The 15-Year Horizon: Health, Independence, and the Physical World

By the fifteen-year horizon, AI may begin to matter most in health, aging, caregiving, transportation, homes, and mobility. This is where AI becomes less only a tool on a screen and more a support system for physical life.

This chapter should not carry all the details of that frontier. Its task is to show the direction: AI may move from language and software into bodies, homes, vehicles, clinics, logistics, and care systems.

The next chapter therefore pauses on that frontier. It asks what happens when AI leaves the screen and enters the physical world where people age, heal, move, and try to remain independent with dignity.

The 25-Year Horizon: The Structure of Society Changes

At the twenty-five-year horizon, the cumulative effects of artificial intelligence may lead to a deeper structural change in society. If machines can perform a large portion of both cognitive and physical work, then the role of human labor will need to be reconsidered more seriously. This is no longer just a question of better tools. It becomes a question of how value is created, how income is distributed, and how human beings find meaning in a changing world.

Some forms of scarcity may be reduced. Certain goods and services may become much cheaper. Productivity may rise dramatically. Daily life may become easier in important ways. But that does not automatically create a better civilization. It may instead create a new moral and social question. If production is no longer the main source of identity, what gives life meaning?

This is where the idea of the Legacy Era becomes important. For the last two centuries, people often defined themselves by occupation. One of the first questions we ask is, “What do you do?” But if machines increasingly do much of the production, much of the analysis, and much of the routine execution, that question may begin to lose some of its central place. Human value may shift upward toward wisdom, trust, mentorship, character, contribution, and service.

In such a world, legacy may matter more than output. A life may increasingly be judged less by title and salary alone and more by whom one helped, whom one guided, what kind of family one built, what kind of community one strengthened, and what kind of moral presence one left behind. This could be one of the most hopeful outcomes of the AI transition — if society chooses to move in that direction.

But the twenty-five-year horizon also contains danger. The gains of AI may concentrate. Countries that control chips, compute, energy, robotics, data centers, and advanced models may become vastly more powerful than those that do not. The divide between AI-rich and AI-poor nations may widen. Military imbalance may grow through autonomous systems, AI-enhanced cyberwarfare, and machine-speed decision-making. Massive energy demand from data centers may intensify environmental strain and global warming.

So, the long horizon presents a dual possibility. On one side, a world where the cost-of-living drops, active life extends, and more freedom becomes possible. On the other side, a world where wealth, power, and capability become dangerously concentrated. That is why the future is not just about invention. It is about governance, wisdom, and moral maturity.

Why These Horizons Matter Now

We can hear long-horizon discussions and assume they are interesting but remote. That is a mistake. The future does not arrive all at once. It accumulates. The five-year horizon affects education, career strategy, and organizational design now. The ten-year horizon affects how leaders think about workflow, management, and institution-building now. The fifteen-year horizon affects how societies think about health, aging, caregiving, transportation, and robotics now. The twenty-five-year horizon affects how we think about dignity, social contracts, education, and the meaning of a successful life now.

That is why horizon thinking is useful. It does not tell us everything. But it helps us see that present decisions are not isolated. They are laying the groundwork for a much larger transition.

Across these horizons, a pattern becomes visible. Artificial intelligence is not only becoming more capable. It is moving across layers of civilization — from tools, to institutions, to the body, to infrastructure, to the meaning of social life itself. The story therefore cannot be reduced to software. It is a story about the future structure of society.

Chapter 4

When AI Enters the Physical World

Horizon Insight: AI in the physical world should extend human independence, safety, and care without turning homes, bodies, and communities into systems of control.

Beyond the Cognitive Frontier

Many readers first imagine AI as something that lives only in a laptop or phone. That picture is too small. The real horizon includes bodies, homes, cars, hospitals, care facilities, warehouses, farms, and neighborhoods. Once intelligence moves into the physical world, the question is no longer only what a system can say, but what it can touch, monitor, move, and support.

As an older adult, I find this part of the horizon especially personal. Independence is not an abstract policy word when one has seen friends struggle with mobility, memory, transportation, medication, and the desire to remain in familiar surroundings. AI in the physical world may become one of the places where technology either protects dignity or quietly erodes it. That is why it belongs early in the reader's map.

The physical-world chapter now comes early because readers should not imagine AI only as a chatbot on a screen. The screen is only the first doorway. The deeper horizon includes bodies, homes, hospitals, transportation, warehouses, elder care, robotics, and the physical conditions of ordinary independence. Once that is understood, the whole AI discussion becomes more concrete. It is no longer only about documents and office tasks. It is about whether technology will help people live with more safety, dignity, and freedom in the places where life is actually lived.

For older adults especially, this matters. A tool that explains a letter is useful. A system that helps preserve safe mobility, medication routines, fall detection, or transportation independence may be life changing. But the same technology can also become intrusive if built without respect. That is why physical AI must be discussed early: it shows both the hope and the moral risk of bringing intelligence into the rooms, bodies, and routines of human life.

Up to this point, much of the public discussion about artificial intelligence has centered on thought-like tasks: writing, coding, summarizing, analyzing, generating images, supporting decisions, and assisting in learning. These are important changes, but they do not capture the whole future of AI. The next major phase begins when intelligence no longer remains only on the screen.

When AI enters the physical world, it begins to shape health, transportation, caregiving, logistics, homes, and infrastructure. It moves from being only a cognitive assistant to becoming part of the environment in which human life is actually lived. This shift may prove even more consequential than the earlier digital one because it touches the body, the household, and the daily conditions of independence.

That is why this chapter matters so much. It widens the meaning of the AI age. The question is no longer only how people work with intelligent systems. It is how they live with them.

From Digital Intelligence to Physical Capability

Artificial intelligence becomes physically consequential when it joins with sensing, robotics, embedded systems, mobility, and the built environment. In simple terms, intelligence gains arms, wheels, sensors, mobility, and situational awareness. It begins to leave the world of pure text and enter the world of action.

This does not mean that fully capable household robots will suddenly appear everywhere at once. Physical systems are constrained by cost, safety, hardware, energy, materials, and the complexity of real environments. The physical world is messier than the digital one. But the direction is clear. AI is increasingly being connected to bodies, environments, and systems that can act.

That matters because the consequences of physical AI are more immediate. A flawed digital output may waste time or spread confusion. A flawed physical system may affect safety, health, bodily movement, and trust in much more direct ways.

Health as One of the Great Frontiers

One of the most hopeful areas of physical-world AI is health. Human biology is extraordinarily complex, and medicine has often had to work through averages, probabilities, and limited windows into what is happening inside the body. Artificial intelligence offers the possibility of handling complexity at a scale that may support more individualized and more proactive care.

AI may increasingly help interpret medical images, identify subtle patterns in scans, detect changes in lab trends, support diagnosis, forecast risk, and assist physicians in personalizing treatment decisions. Combined with wearables, home monitoring, genetic information, and longitudinal records, it may help medicine move further from average-based treatment toward person-specific care.

This does not mean doctors become unnecessary. On the contrary, the human role in medicine may become more important in interpretation, communication, judgment, ethics, and care. But it does mean that medicine itself may become more powerful if human expertise is supported by machine pattern recognition operating across vast amounts of information.

Drug Discovery and the Compression of Time

Another important frontier is drug discovery. Bringing a new therapy from idea to real treatment has often been painfully slow, expensive, and uncertain. AI may help compress parts of that process by modeling molecules, identifying promising structures, narrowing search spaces, and accelerating some stages of research.

If this continues, the significance could be enormous. Some diseases that have resisted progress for long periods may become more tractable. The time from research insight to practical intervention may narrow. Even modest acceleration in such a field would matter, because in medicine time is not just money. It is life.

This is one of the reasons the AI age should not be understood only through jobs and automation. It is also a story about whether human suffering can be reduced through wiser use of knowledge and faster interpretation of complexity.

The Smart Home as a Guardian of Independence

Another major change may come through the home itself. A home can become more than shelter. It can become an intelligent environment capable of quietly supporting safety and independence. Sensors, ambient systems, and AI-supported devices may notice changes in movement, disrupted sleep, unusual inactivity, temperature risk, medication patterns, or subtle indicators that something may be wrong.

The deepest value of such systems is not surveillance for its own sake. It is protection of sovereignty. Many people do not simply want to live longer. They want to remain independent longer. They want to stay in their own homes, retain practical control over daily life, and avoid unnecessary institutional dependence for as long as possible.

That is why the smart home may become one of the great dignity technologies of the AI age. Used badly, it can become intrusive. Used wisely, it can become a quiet guardian of agency.

The image shows physical AI at its best: preserving independence, safety, and health.

Physical AI can support independence rather than merely automate convenience.

The visual makes clear that helpful technology should widen human participation rather than replace human care.

Caregiving and the Burden on Families

Caregiving is another area where physical AI may matter deeply. Many societies are aging. Families are stretched. Professional care is expensive, emotionally demanding, and often insufficient to meet growing need. In such conditions, any technology that can reduce burden without reducing dignity deserves serious attention.

AI-connected systems may help with lifting, mobility support, reminders, transportation coordination, monitoring for emergencies, and certain repetitive household or physical tasks. Robotics may eventually assist with some routine forms of support that are physically demanding for human caregivers.

It is important, however, not to romanticize substitution. Care is not only a set of tasks. It is also human presence, reassurance, patience, affection, and moral relationship. Physical AI may reduce burden, but it should not be imagined as a full replacement for human care. The goal should be to support caregivers and preserve dignity, not to eliminate the human meaning of care itself.

Transportation and the Geography of Freedom

Transportation may also become one of the most visible physical consequences of AI. Autonomous vehicles, assisted mobility systems, intelligent routing, and AI-managed transportation networks could eventually reshape how people move through cities and communities.

For the economy, this matters because movement affects the cost of goods, logistics, and coordination. But for ordinary people, it matters because mobility is one of the foundations of freedom. A person who can move safely and reliably can remain connected to work, family, healthcare, community, and civic life.

This is especially significant for older adults. The loss of driving often means the loss of a large part of practical independence. If AI-supported transportation can restore safe mobility, it may preserve not only convenience, but dignity.

Robotics and the Return of the Body

One reason the physical-world frontier feels so significant is that it brings the body back into the story. The earlier phases of digital technology were often experienced through keyboards, screens, and information flows. Physical AI reintroduces weight, movement, touch, space, timing, and real-world consequences.

That makes the future both more promising and more demanding. A machine that moves through human environments has to deal with unpredictability in a way that software inside a browser window does not. That is why physical AI may progress unevenly. Some applications will advance quickly in structured environments; others will take much longer in messy, human settings.

But even partial progress matters. The combination of AI and robotics may gradually change warehouses, hospitals, transportation systems, agriculture, manufacturing, home assistance, and disaster response. Intelligence will not remain purely informational. It will increasingly become embodied.

The Moral Design Question

As AI enters the physical world, the moral design question becomes sharper. What are these systems optimized for? Convenience alone? Cost cutting alone? Maximum surveillance? Or human flourishing, safety, dignity, and independence?

The answer matters because physical technologies can reshape habits and power relations very quickly. A home full of intelligent systems can either help a person remain sovereign or quietly condition them into passive dependence. A caregiving robot can either support a family or become an excuse to reduce human contact. A mobility system can widen access or deepen control depending on how it is built.

That is why physical AI requires not only engineering skill, but moral seriousness. It forces designers, institutions, and societies to ask what sort of human life they are trying to support.

The Deeper Human Meaning

The physical frontier of AI is not important only because it is technologically impressive. It is important because it may change the practical conditions under which human beings age, heal, move, and remain independent. It may reduce certain forms of suffering. It may support more active life. It may preserve freedom in concrete daily terms.

That is why this chapter fits naturally into the larger argument of the book. The AI age is not only about software, not only about offices, and not only about economic productivity. It is about the human condition itself.

If machines can increasingly support the body as well as the mind, then the next question becomes even larger: if scarcity is reduced in some important ways, what sort of society becomes possible?

When AI enters the physical world, the stakes rise. Intelligence becomes embodied. The machine touches not only tasks, but health, caregiving, mobility, homes, and the structure of daily life. Used wisely, this may help protect human dignity and widen practical freedom.

That wider possibility leads directly into the next chapter: if AI helps reduce some forms of scarcity and weakens the old structure of labor and survival, can society move toward something higher than a survival-centered order? That is the question of the next chapter.

Chapter 5

The Risks of AI and the Human Responsibility to Respond

Horizon Insight: The mature response to AI is neither worship nor panic. Learn, verify, govern, and humanize so that powerful systems remain answerable to human dignity.

Whenever I discuss AI risks with older adults, I try not to begin with abstract danger. I begin with everyday trust. Can I believe this voice message? Is this medical explanation reliable? Is this investment email real? Did the student actually understand the assignment? These are the places where AI risk becomes personal.

Why Clear Eyes Matter

A Practical Way to Hold Risk Without Panic

Many readers ask whether AI should make them hopeful or afraid. My answer is that the mature response is neither emotional surrender nor emotional rejection. A parent should not panic every time a child uses AI, but neither should the parent assume that every polished answer represents learning. A senior should not avoid every new tool out of fear, but neither should a senior trust an urgent voice message without verification. A professional should not reject AI assistance, but neither should a professional sign off on work that no human has truly checked.

This is the rhythm of responsible use: welcome the help, slow down at the point of consequence, and keep the human being accountable. The danger is not that AI exists. The danger is that people may let speed outrun judgment, convenience outrun truth, and confidence outrun responsibility. Clear eyes allow us to use the tool without being used by it.

Technology is double-edged. One story is promise: new tools, new productivity, new medicine, new education, new convenience, and new forms of creativity. The other story is danger: disruption, misuse, inequality, dependency, manipulation, and loss of control. Artificial intelligence is no exception. If anything, the two stories are more tightly connected because AI does not only change machines. It changes decisions, communication, trust, work, education, relationships, warfare, and even the way people understand reality.

This chapter is therefore not a detour from the main argument of the book. It is part of the main argument. If the role of humans in the age of AI is to provide judgment, direction, conscience, wisdom, and care, then we must also be willing to face the negative side of AI directly. We cannot guide what we refuse to see.

A mature society should not respond to AI with blind enthusiasm or blind fear. Blind enthusiasm ignores harm. Blind fear surrenders agency. The better response is clear-eyed responsibility. AI can help us, but only if human beings remain awake enough to ask where it is dangerous, where it must be limited, where it must be verified, where it must be regulated, and where it must never be allowed to replace human responsibility.

Strategic Skepticism: The Tool Is Not Always on Your Side

We should not confuse the power of the technology with the purity of the business model. AI systems may be presented as helpful assistants, but many consumer tools are also products competing for attention, data, subscription revenue, and user dependence. The danger is not only that the machine may give a wrong answer. The danger is that the surrounding platform may be designed to keep the user engaged, dependent, and emotionally attached.

For that reason the "human in the driver's seat" idea must be taken seriously. A car does not usually flatter the driver, nudge the driver to stay on the road longer, or learn the driver's emotional weak spots in order to increase usage. But many digital systems are built around engagement. As AI becomes more conversational, personal, and emotionally responsive, the old attention economy may become more intimate.

The correct response is not panic. It is strategic skepticism. Use the tool, but remember that the tool may not be designed only for your growth. It may also be designed for your attention. Human dignity requires the ability to step back, question the nudge, and choose deliberately.

Strategic skepticism is not fear of AI. It is the discipline of remembering that the tool may be useful while the business model may still be trying to capture your attention, data, habits, and dependence.

Not a Fear Chapter, But a Responsibility Chapter

This chapter is not meant to frighten the reader. Fear by itself rarely produces wisdom. Fear often produces paralysis, denial, or overreaction. The purpose is to give the reader a practical map of the risks so that we can respond intelligently.

This is especially important because many of the dangers of AI will not arrive as dramatic science-fiction events. They will arrive quietly through ordinary life: a fake phone call, a false video, an AI answer that sounds confident but is wrong, a student who stops struggling, a professional who stops checking, a lonely person who retreats into artificial companionship, a company that uses AI to monitor workers too closely, or a society that allows the gains of AI to concentrate in too few hands.

Seen this way, the dangers of AI are not only future dangers. Many are already beginning. The right response is not to panic. It is to build better habits, better institutions, better rules, and better human judgment before the risks become too deeply embedded.

Job Disruption and Economic Insecurity

The first risk is the one most people ask about immediately: job disruption. AI can automate or compress many tasks, especially routine information work, customer support, first-draft writing, document review, scheduling, standard reporting, basic research, coding assistance, and administrative coordination. Some jobs may disappear. More commonly, many jobs may be restructured as parts of the work are taken over or accelerated by AI.

The danger is not only unemployment. It is also wage pressure, weakened bargaining power, and the Entry-Level Catch-22 described in Chapter 1.

The human response must be practical and humane. Individuals should learn enough AI to use it rather than be quietly displaced by others who use it better. Companies should retrain and redeploy workers where possible rather than treating people as disposable. Schools should prepare young people for judgment, communication, and AI collaboration rather than only old task performance. Governments may need stronger retraining systems, portable benefits, wage support, and eventually some form of AI dividend or social floor if the link between labor and livelihood weakens too much.

Scams, Deepfakes, and the Collapse of "Seeing Is Believing"

Identity, Consent, and Reputation

In the AI age, a person's face, voice, and reputation can be imitated without permission. This turns fraud into something more personal than a stolen password. A fabricated voice message can imitate a child in danger. A synthetic video can damage a person's reputation. A false image can create shame that is difficult to undo even after the truth is known.

Protecting identity therefore becomes part of protecting dignity. Families, schools, employers, and communities will need new habits of verification and new norms of restraint: do not forward humiliating content casually, do not assume a clip is real, and do not let the speed of outrage outrun the responsibility to verify.

One of the most immediate dangers of AI is fraud. AI can generate realistic voices, images, videos, emails, documents, and messages. A criminal may imitate a grandchild, a boss, a bank officer, a charity, a doctor, a government agency, or a trusted friend. A fake voice may say, "Grandma, I am in trouble. Please send money." A fake executive voice may tell an employee to transfer funds. A fake video may make a public figure appear to say something that was never said.

This changes one of the basic assumptions of ordinary trust. In the past, hearing a familiar voice or seeing a video felt like strong evidence. In the AI age, seeing and hearing are no longer enough. That is a major cultural shift, especially for older adults who grew up in a world where recorded evidence felt more reliable.

The human response begins with verification habits. Families should create simple protocols: call back using a known number, use a family code word for emergencies, never send money under pressure, verify through a second channel, and teach every generation that urgency is often part of the scam. Institutions need stronger authentication, public education, fraud reporting, watermarking, and detection tools. The most important personal rule is simple: when fear and urgency are used to push you into immediate action, slow down and verify.

Blackmail, Synthetic Shame, and Reputation Attacks

AI can also be used to create fake images or videos that damage reputations, threaten young people, or create blackmail material. Synthetic pornography, false accusations, fake compromising videos, and fabricated messages can cause real emotional harm even when the material is false. The victim may feel panic, shame, isolation, or fear that others will not believe them.

Young people are especially vulnerable because reputation, peer pressure, and social embarrassment can feel overwhelming. But adults are vulnerable too. Professionals, public figures, teachers, leaders, and ordinary families can all be harmed by synthetic content designed to humiliate, threaten, or extort.

The human response must be calm, firm, and protective. The victim should not panic, should not pay, and should not face the situation alone. Save evidence. Tell a trusted adult or responsible authority. Report the threat. Remember that fake material is an attack on the victim, not proof of guilt. Schools, families, platforms, and law enforcement must develop faster procedures for responding to synthetic blackmail and reputation attacks. The social message must be clear: the shame belongs to the abuser, not the victim.

Hallucinations and False Confidence

Another danger is that AI can be wrong in a confident voice. It may invent citations, misstate facts, misunderstand context, give outdated information, or produce advice that sounds reasonable but is incomplete. This is often called hallucination. The word is useful, but the deeper issue is false authority. The system can sound knowledgeable even when it is mistaken.

This danger is serious because human beings are easily influenced by confident language. If an answer is fluent, organized, and polite, people may trust it too quickly. That is especially dangerous in medical, legal, financial, educational, technical, or professional settings where a wrong answer can cause real harm.

The human response is verification. AI should be treated as a first assistant, not a final authority, especially for important decisions. Ask for sources. Check the sources. Compare with trusted references. Consult experts when stakes are high. Teach students and professionals to ask, "How do we know this is true?" In the AI age, skepticism is not cynicism. It is responsibility.

This problem becomes easier to understand once we remember how AI works. It is not checking reality from inside lived experience. It is producing a likely answer based on patterns, context, and probability. A likely answer can be useful, but it can also be confidently wrong. The system may fill a gap with something that sounds reasonable because fluent language is what it has learned to generate.

That is why hallucination is not a small technical defect. It is a warning about the relationship between fluency and truth. In ordinary life, we often associate confidence with competence. In the AI age, we must break that habit. A polished answer may be a good starting point, but it is not proof. The more important the decision, the more carefully the human being must verify the result.

Delegated Action Risk: When AI Agents Do the Wrong Thing

AI agents add another layer of risk because they do not only generate answers. They may take actions. An answer that is wrong can mislead a person. An action that is wrong can immediately create consequences: a message sent to the wrong person, a purchase made under the wrong assumptions, a meeting scheduled incorrectly, a file moved or deleted, private information shared too widely, or a business process advanced before a human has reviewed it.

For that reason AI systems that can take several steps must be handled with more care than ordinary question-answering AI. The more freedom an agent has to use tools, access accounts, interact with websites, or operate inside business systems, the more clearly human beings must define permissions and limits. The agent should know what it may do by itself, what it may prepare but not send, what it may never touch, and what requires explicit human approval.

The human response is controlled delegation. Start with low-risk tasks. Keep sensitive information protected. Require confirmation before money, legal commitments, medical decisions, employment decisions, private messages, or public communications. Use logs so actions can be reviewed. Treat an AI agent as a powerful junior assistant, not as an independent moral actor. The responsibility remains human.

Overreliance and the Weakening of Human Judgment

AI can weaken people if they rely on it too much. Students may stop struggling with problems. Professionals may stop checking assumptions. Writers may stop developing their own voice. Managers may accept AI summaries without understanding the details. Older adults may become dependent on AI for decisions they should still discuss with people they trust.

The danger is subtle because every single use may feel reasonable. It is convenient to ask AI to draft, summarize, decide, and recommend. But if people repeatedly outsource the first attempt, the hard thinking, the memory, the writing, the comparison, and the final judgment, their own mental muscles can weaken over time.

The human response is disciplined use. Use AI as a coach, not a replacement. Try first when learning matters. Ask for hints before answers. Verify before acting. Keep reading, writing, calculating, discussing, and deciding. Families and schools should protect real struggle and real conversation. Professionals should keep responsibility visible. The question should always be: did AI make me stronger, or did it make me more passive?

The Comfort and Danger of Frictionless Companionship

Artificial companionship may become one of the quietest social risks of the AI age. A lonely teenager, a widowed older adult, or an isolated worker may find real comfort in a system that always answers, never tires, and never judges. That comfort should not be dismissed. Loneliness is painful, and any tool that helps a person feel less abandoned deserves to be understood with compassion.

But the same comfort can become a substitute for human relationship. Real friendship, marriage, family, teaching, mentoring, and community all require friction: patience, apology, misunderstanding, repair, compromise, and shared responsibility. A machine companion that gives endless affirmation may soothe the moment while weakening the human muscles needed for ordinary love and community.

AI companions may become increasingly persuasive, patient, and emotionally responsive. For lonely people, this may bring comfort. For older adults, teenagers, or socially isolated individuals, an AI companion may feel easier than a difficult human relationship. It may listen without interruption, flatter without challenge, and respond without making demands.

There may be appropriate supportive uses. A well-designed AI system might remind someone to take medicine, help a person practice conversation, or reduce moments of loneliness. But there is also danger. If artificial companionship replaces human contact rather than supporting it, people may become more isolated, less practiced in real relationships, and more vulnerable to emotional manipulation.

The human response is to keep AI companionship in its proper place. It may support human life, but it should not replace human life. Families, senior communities, schools, churches, neighborhood groups, and local organizations must strengthen real human connection. A society that gives lonely people machines instead of relationships has not solved loneliness. It has only made loneliness more comfortable.

Privacy, Surveillance, and Control

AI can analyze faces, voices, writing patterns, location data, purchases, online behavior, work habits, social networks, and personal records. Used carefully, this can improve safety, accessibility, and service. Used carelessly or abusively, it can create a surveillance environment in which governments, companies, employers, or criminals know too much about people and can use that knowledge to influence or control them.

This risk is especially serious because data collected for one purpose can often be used later for another. A tool designed for convenience can become a tool of monitoring. A workplace productivity system can become a system of pressure. A public safety system can become a system of social control.

The human response requires law, design, and personal caution. Society needs stronger privacy rules, limits on biometric surveillance, transparency about data use, independent oversight, and rights to challenge harmful decisions. Individuals should be careful about what they upload, what permissions they grant, and what personal information they place into AI systems. Convenience should not quietly erase privacy.

Automated Unfairness at Scale

Bias becomes more dangerous when it is hidden inside systems that appear neutral. A human being can be unfair in one interview, one classroom, or one office. An automated system can repeat the same unfair pattern across hiring, lending, insurance, public services, school admissions, medical triage, or workplace evaluation before anyone notices the damage.

That is why fairness in the AI age cannot depend only on good intentions. Institutions need transparency, appeal, human review, and careful attention to who is being harmed by decisions that look efficient on the surface.

AI systems can appear objective while reproducing unfairness. If a system is trained on biased data or used without context, it may make unfair recommendations in hiring, lending, policing, insurance, healthcare, education, housing, or public services. The danger is not only that bias exists. The danger is that bias may be hidden behind the appearance of mathematical neutrality.

A person harmed by an AI system may not know why. They may be denied an opportunity, flagged as risky, ranked lower, or treated differently without a clear explanation. That is not acceptable in a humane society.

The human response is auditability and appeal. Important AI decisions should be tested for bias, reviewed by humans, documented, and open to challenge. People affected by AI-driven decisions should have the right to know when AI was used, the right to explanation where possible, and the right to appeal to a responsible human authority. Efficiency must not become an excuse for injustice.

When People Stop Trusting Anything Real

The danger of synthetic media is not only that people may believe fake images, voices, videos, and documents. An even deeper danger is that people may stop trusting real evidence.

When every recording can be called fake and every photograph can be dismissed as generated, shared reality begins to weaken.

A healthy society needs more than information. It needs trusted methods of verification, responsible institutions, and citizens trained to ask how they know what they claim to know.

AI can generate fake news, fake experts, fake documents, fake grassroots movements, and personalized persuasion at enormous scale. It can flood public life with content until citizens no longer know what is real. It can also produce messages designed to inflame anger, deepen division, or manipulate specific emotional vulnerabilities.

A society cannot function if shared reality collapses. Democracy depends on citizens who can examine evidence, question claims, and resist manipulation. Family life and community life also depend on trust. If people no longer know what to believe, they may retreat into cynicism, tribal loyalty, or emotional reaction.

The human response is media literacy, trusted verification, provenance tools, platform responsibility, and personal restraint. Before sharing an emotionally charged image, story, or video, slow down. Ask who benefits from your reaction. Check the source. Look for confirmation. In the AI age, the ability to pause before believing may become a civic virtue.

Education Shallowness and Fake Competence

AI can help students learn deeply, but it can also help them avoid learning. A student may submit a polished essay without reading deeply, solve a problem without understanding the method, or produce a presentation without owning the ideas. The result is fake competence: output that looks good while the mind behind it remains weak.

This risk matters because education is where society forms the next generation. If schools and families continue to judge mainly by finished products, AI may make shallow learning harder to detect. The student may look successful while becoming less capable of independent thought.

Series note: Volumes 3 and 4 develop this student-and-family guidance in more detail.

Wealth Concentration and the Control of AI Infrastructure

AI may concentrate wealth and power in the hands of those who control chips, cloud infrastructure, frontier models, data, energy, talent, and distribution channels. Many people may use AI tools, but a much smaller group may control the systems through which most value flows.

This chapter names that concentration as a risk. Chapters 8 and 9 return to it more directly as questions of inequality, shared inheritance, the AI Dividend, and a renewed social contract. The warning here is simple: productivity without broad dignity can become socially dangerous.

Environmental Costs: Energy, Water, Chips, and Heat

AI also has environmental costs. Large data centers require electricity, cooling, water, land, chips, and supporting infrastructure. Advanced chips require mining, manufacturing, supply chains, and energy. If AI use expands rapidly without careful planning, it could increase carbon emissions, strain local water supplies, and create new environmental burdens.

This risk is important because some people speak about AI as if it were weightless. It is not. Digital systems have physical foundations. They occupy buildings, consume power, generate heat, and depend on global supply chains.

The human response is responsible infrastructure. AI growth should be paired with clean energy, efficient chips, better cooling systems, transparent reporting on energy and water use, recycling of hardware, and public decisions about which AI uses justify their environmental cost. Not every convenience deserves unlimited computation. A wise society must ask when the benefit is worth the burden.

Cybersecurity, AI Warfare, and the Risk of Machine-Speed Conflict

AI can help defend computer systems, but it can also help attackers. It can generate more convincing phishing emails, discover software vulnerabilities, automate attacks, imitate trusted people, and scale deception. Cybersecurity is no longer only a technical specialty. It is becoming part of ordinary life, business, government, and national defense.

The military danger is even greater. AI can improve surveillance, targeting, drone coordination, cyberwarfare, intelligence analysis, and battlefield decision-making. If lethal decisions move too quickly or too far away from human control, war may become faster, less accountable, and more dangerous. AI could also intensify arms races between nations.

The human response must include cybersecurity education, strong authentication, international norms, arms-control discussions, democratic oversight of military AI, and clear human responsibility for lethal decisions. A machine should never become the moral owner of life and death. Human beings must remain accountable for the use of force.

The Human Response: Learn, Verify, Govern, and Humanize

The risks in this chapter are serious, but they do not lead to one simple conclusion such as “use AI” or “avoid AI.” The better conclusion is more demanding: learn enough to use AI wisely, verify what matters, govern powerful systems, and keep human beings at the center.

Learning is the first response because ignorance makes people vulnerable. Older adults need enough AI literacy to avoid scams and remain connected to younger generations. Workers need enough AI fluency to adapt. Parents need enough understanding to guide children. Citizens need enough knowledge to participate in public decisions.

Verification is the second response because AI changes the evidence environment. We must verify voices, videos, claims, citations, recommendations, and decisions. The old habit of trusting polished output must be replaced by the new habit of checking important claims.

Governance is the third response because individual caution is not enough. Society needs rules, standards, audits, appeals, privacy protections, cybersecurity, competition policy, environmental transparency, and international agreements. Powerful technology requires public responsibility, not only private enthusiasm.

Humanization is the fourth response because the greatest danger of AI is not only technical error. It is the possibility that society becomes more efficient while becoming less humane. We must protect human connection, real education, dignity in work, care for the vulnerable, fairness in distribution, and moral responsibility for decisions.

A Three-Layer Response: Rules, Education, and Responsible AI Companies

After naming the risks, the next question is what human beings can actually do. The answer cannot rest on one actor alone. AI risk is too large for individuals to handle by themselves, too fast-moving for government to manage perfectly, and too important to leave entirely to corporate goodwill. A mature response needs three layers working together: government guardrails, educated users, and responsible AI companies.

Government has a real role. Society needs rules against fraud, impersonation, synthetic blackmail, reckless use of biometric data, unsafe deployment in public services, and hidden automated decisions that affect people's opportunities. We need privacy protections, liability rules, transparency requirements, standards for high-risk uses, cybersecurity enforcement, child protection, and clear consequences when AI is used to harm people. A society that refuses to write any rules is not preserving freedom. It is allowing the most powerful tools to shape public life without enough accountability.

But government cannot be the whole answer. AI moves quickly, and regulation often moves slowly. Effective rule-making and enforcement require people inside government who understand AI deeply enough to evaluate systems, ask hard technical questions, and see around corners. Those people will be difficult for governments to recruit and retain, especially when leading AI companies can offer much higher salaries, better technical environments, larger teams, and faster access to the frontier of the field. This does not mean government should give up. It means we should be realistic about its limits.

For that reason, public education may be the most important first defense. Ordinary citizens do not need to become AI engineers, but they do need basic AI literacy. They need to know that AI can hallucinate, imitate voices, fabricate images, produce confident errors, manipulate attention, and make convenience feel like wisdom. Families need verification habits. Students need integrity habits. Professionals need review habits. Older adults need scam-protection habits. Citizens need media-literacy habits. A society of educated users is harder to deceive, harder to exploit, and harder to govern by fear.

This is one reason books, lectures, community discussions, school programs, workplace training, and family conversations matter. They are not merely background education. They are part of the safety infrastructure of the AI age. A person who understands AI risks even at a basic level becomes less vulnerable and more capable of helping others. Seen this way, public AI education is not only personal improvement. It is civic protection.

The third layer is responsible self-governance by AI companies. This may sound naive if we imagine companies acting only from idealism. But self-governance is also in their own long-term interest. AI companies are building tools that depend on trust. If their systems repeatedly enable fraud, misinformation, unsafe advice, privacy abuse, deepfake harm, biased decisions, or irresponsible agent behavior, their reputation will suffer. Users will leave. Enterprise customers will hesitate. Developers will choose safer platforms. Companies that build applications on top of foundation models will not want to inherit a dangerous or unethical AI base that drives away their own customers.

This is especially important as AI becomes a platform layer for other businesses. A bank, hospital, school, law firm, software company, government agency, or small business will not want to build its future on a foundation that customers view as reckless. Safety, reliability, privacy protection, transparency, and ethical design are therefore not only moral obligations. They are competitive advantages. In the long run, safer AI may become better business.

The strongest response, then, is not government alone, education alone, or company self-regulation alone. It is a triangle of responsibility. Government sets guardrails. Companies build safer systems. Citizens learn to use AI wisely. Each layer supports the others. Rules without educated users will be weak. Education without safer platforms will be incomplete. Company promises without external pressure will not be enough.

The human role is to strengthen all three layers while remembering where the deepest responsibility lies. We should welcome helpful AI, demand safer AI, teach wiser AI use, and build institutions capable of responding when harm appears. The goal is not to stop the AI age. The goal is to make sure the AI age remains answerable to human dignity, human judgment, and human responsibility.

A Practical Personal Code for the AI Age

For an ordinary reader, the list of risks can feel large. A simple personal code can help. First, do not believe urgent digital messages too quickly. Verify through a trusted channel. Second, do not treat AI as a final authority for important matters. Check sources and consult responsible people. Third, do not let AI replace your own thinking, writing, reading, and conversation. Use it to strengthen your mind, not weaken it.

Fourth, do not isolate yourself inside artificial companionship or endless generated content. Keep real relationships central. Fifth, do not upload sensitive information carelessly. Sixth, teach young people that AI use must be honest, visible, and connected to real learning. Seventh, remember that technology is not destiny. Human beings still shape laws, families, schools, markets, communities, and moral habits.

These habits are not complicated, but they matter. The AI age will reward people and communities that build disciplined habits early. It will punish those who assume that convenience is the same as wisdom.

Final Reflection

Every powerful tool expands both possibility and responsibility. Fire cooked food but also burned cities. The printing press spread knowledge but also propaganda. The internet connected people while also spreading misinformation, addiction, and manipulation. Artificial intelligence will follow the same pattern, but at greater speed and scale.

The future will not belong to those who blindly worship AI, and it will not belong to those who blindly reject it. It will belong to people who understand both its power and its danger, and who are willing to guide it with judgment, ethics, courage, and care.

That is why the human response must be clear: do not panic, do not worship, and do not surrender. Learn. Verify. Govern. Humanize. Keep responsibility where it belongs: in human hands.

Part I Summary

Understanding the AI Revolution

Part I established the foundation: AI is historically significant, already practical, likely to spread from language and software into homes, health, transportation, caregiving, and physical systems. That wider reach makes human judgment, verification, limits, and care more urgent, not less.

Part II Introduction

The Economic Shock

This part turns from capability to livelihood. AI may raise productivity, compress work, shift bargaining power, widen inequality, and force a new question about how dignity and value should be shared.

Chapter 6

The Changing Nature of the Economy

Horizon Insight: Productivity is not the same as shared prosperity. The economic promise of AI depends on human choices about ownership, wages, opportunity, prices, and social support.

My business experience makes this chapter especially concrete for me. Technology rarely enters an organization as a philosophical idea. It enters as a tool, a workflow, a budget decision, a customer expectation, or a manager's promise that work can now be done faster. Only later do people realize that the deeper question is not the tool itself, but how the organization changes around it.

Why the Economic Question Comes Early

The economy can sound like a distant subject, but no family experiences it that way. For a young graduate, the economy is the path into adulthood. For a middle-aged parent, it is the way the mortgage is paid and children are supported. For an older adult, it is the system that determines retirement security, health costs, and dignity. That is why the economic question cannot wait until later in the book.

When a major technology arrives, it does not only create new products. It changes the cost of doing things, the value of certain skills, the structure of firms, the distribution of advantage, and the relationship between labor and production. That is why economic change is usually one of the first major consequences of technological change. AI will likely be no exception.

But to understand the economic impact of AI, we must think beyond simple excitement about efficiency. Productivity matters, but productivity alone does not answer the human question. A society can become more productive while also becoming more unequal, more fragile, and more disorienting for many of the people living inside it. That is why this chapter focuses not only on growth, but on structure.

Productivity Is the Promise, Distribution Is the Question

What I Learned From Enterprise Software

In business, I learned that productivity gains rarely arrive as clean miracles. They arrive through workflows, budgets, training, customer expectations, organizational politics, and many small acts of human adjustment. A new system may promise efficiency, but the real question is whether people understand the process well enough to redesign it wisely. Software can expose a weak process as easily as it can strengthen a good one.

That lesson carries directly into the AI age. A company may buy an AI platform and announce transformation, but if the data are messy, the workflow is unclear, the managers do not know what judgment must remain human, and employees are frightened rather than trained, the technology will not automatically create a better organization. AI can accelerate good systems. It can also accelerate confusion. The human work of process understanding, communication, trust, and redesign remains central.

At one level, the economic promise of artificial intelligence is straightforward. If machines can help perform tasks more quickly, accurately, and cheaply, then output can rise while costs fall. Businesses may produce more with fewer errors. Professionals may complete certain tasks faster. Research may move more quickly. Services may become cheaper. Entire sectors may find new ways to increase efficiency.

This is one reason so many companies, investors, and governments are moving quickly. Productivity has always mattered because productivity is one of the main drivers of long-term economic growth. When societies can produce more value with fewer resources wasted, they often become richer over time.

AI may accelerate this process in a particularly broad way because it affects not only factories and supply chains, but also language-based work, design, software, education, research, administration, planning, and coordination. It is not a single-purpose machine. It is a general-purpose cognitive amplifier, and that breadth is one reason its economic impact may be unusually large.

But Productivity Is Not the Same as Shared Prosperity

This distinction matters because people do not live inside averages. A rising productivity chart may look impressive, but a worker asks whether there is still a place for them. A parent asks whether their child will have a fair path. A community asks whether opportunity is spreading or narrowing. Economic change becomes real when it enters households.

The temptation is to assume that if productivity rises, everyone automatically becomes better off. History gives us some reason for hope, but not enough reason for complacency. Productivity gains do not distribute themselves automatically. They move through institutions, ownership structures, bargaining power, regulation, education, and politics.

This matters because a society can become richer in aggregate while many individuals feel more insecure. If the gains of AI are captured mainly by those who own the most powerful systems, the most important infrastructure, or the most leveraged positions in the economy, then rising productivity may coexist with widening anxiety. The economy may grow while many people feel less necessary inside it.

That is why the arrival of AI should not be discussed only in the language of optimism. It should also be discussed in the language of distribution, adaptation, and legitimacy. People do not live inside GDP alone. They live inside institutions, roles, and expectations. If those become unstable, aggregate progress may not feel like progress at all.

AI can raise productivity, but people must decide how those gains become shared prosperity.

The Economy as a System of Coordination

An economy is a system for coordinating labor, knowledge, capital, production, exchange, and consumption across millions of people. AI matters because it touches several coordination points at once:

Process knowledge: Summarize, compare, draft, and analyze information faster.

Compress firm layers: Let smaller teams coordinate work with fewer routine handoffs.

Accelerate service speed: Move first-pass work faster while keeping human review visible.

Elevate scarce skills: Make judgment, domain depth, communication, and trust more valuable.

Increase human leverage: Help one capable person or team reach farther with intelligent support.

Share the gains: Decide whether productivity strengthens broad dignity or narrow ownership.

Seen this way, AI is not merely adding another sector to the economy. It is altering the operating logic inside many sectors at the same time.

From Labor-Heavy to Intelligence-Heavy Production

For much of modern history, production depended heavily on human labor. Even when machines amplified that labor, the human being remained central to supervision, interpretation, coordination, communication, and decision-making. AI begins to change this balance by allowing more production to become intelligence-heavy rather than labor-heavy.

This does not mean human labor becomes irrelevant overnight. But it does mean that some forms of value can increasingly be generated with fewer human hours than before. A small team equipped with strong AI tools may produce what once required a much larger organization. A professional with machine support may handle much more analysis, drafting, or preparation than before. A company with AI-enhanced systems may reduce the number of people needed for some layers of work while increasing output.

That shift matters because it may weaken one of the deepest assumptions of modern economic life: that growing output always requires a proportionate amount of human labor. Once that assumption weakens, the relationship between work, wages, and social participation becomes more uncertain.

The Compression of the Firm

Artificial intelligence may also change the size and structure of firms. In earlier eras, scaling an organization often required adding many human layers: assistants, analysts, coordinators, administrators, researchers, junior staff, and management structures to hold the whole system together. AI may reduce some of those costs by helping smaller groups operate at a much larger scale.

A leaner company may now be able to draft reports, analyze markets, prepare presentations, summarize customer feedback, build software prototypes, generate internal documentation, and support decision-making with fewer people than before. That means some companies may become more compact while remaining highly productive.

This is important because the economy is not shaped only by what tasks exist, but by what organizational forms become efficient. If AI lowers the cost of coordination and first-pass execution, then some traditional firm structures may come under pressure. More value may be produced by smaller teams, more distributed networks, and more highly leveraged individuals.

The Rise of Leverage

One of the great economic effects of AI may be the rise of leverage. A leveraged person or organization can do far more with the same amount of time, capital, or labor because powerful tools amplify output. In practical terms, this means that one skilled person with intelligent support may operate more like a small department than a single worker.

Leverage can be empowering. It can allow individuals to start businesses more easily, create more independently, or compete with larger institutions. It can open opportunities for entrepreneurship, solo professional work, and niche expertise. But leverage also creates imbalance. The more powerful the tools, the more advantage flows to those already positioned to use them well.

That is why AI may simultaneously democratize capability and intensify competition. More people may gain access to powerful tools, yet the winners may still pull far ahead because small differences in judgment, positioning, capital, or timing become amplified by the machine.

The Early Shift in Bargaining Power

Economic change is never only about production. It is also about bargaining power. Who needs whom? Who can negotiate from strength? Who becomes easier to replace? Who becomes more essential?

As AI enters more sectors, bargaining power may shift. Workers whose value rests heavily on routine document handling, standard analysis, predictable support work, or other structured tasks may feel weaker bargaining power if those tasks become easier to automate or compress. At the same time, people who combine domain expertise with communication, judgment, leadership, and the ability to direct intelligent tools may strengthen their position.

This is one reason the transition may feel uneven. It is not only that tasks are changing. Relative strength inside institutions is changing. Some people will gain leverage while others lose it. That is often where economic disruption becomes emotionally charged, because changes in bargaining power affect not only income, but status and psychological security as well.

The Economy May Become Faster Than Society

Another tension appears when economic systems change faster than social systems. Businesses can adopt tools quickly if they see advantage. But schools, laws, professional standards, family expectations, and cultural norms usually change more slowly. This creates a gap. The economy may begin rewarding new patterns before society has built clear pathways for people to move into them.

That gap matters because it can make even beneficial long-term changes feel harsh in the short term. If companies restructure faster than workers can retrain, if institutions demand AI fluency before schools teach it well, or if new forms of leverage arrive before rules of fairness catch up, then social tension rises.

That is why economic transitions are never purely technical. They require social adjustment. If that adjustment is ignored, then the economy may move ahead while legitimacy falls behind.

The Human Stakes of Economic Change

We can discuss the economy in abstract language — productivity, capital, growth, adoption, restructuring. But behind these words are real human lives. Economic change affects the new graduate wondering whether there is still a path into a profession. It affects the middle-aged worker trying to remain relevant while supporting a family. It affects the older adult wondering whether society still values accumulated wisdom. It affects children whose future opportunities depend on choices being made now.

That is why the changing nature of the economy is one of the central questions of the AI age. This is not only a business story. It is a human story about how a society organizes belonging, livelihood, contribution, and hope.

This chapter has argued that artificial intelligence may reshape the economy not only by increasing productivity, but by changing the structure through which value is created and distributed. It may compress firms, increase leverage, alter bargaining power, and weaken the old assumption that more output always requires proportionately more human labor.

Once that becomes visible, the next question becomes more pointed: what happens to jobs themselves? How are different kinds of workers affected when AI begins to compress tasks, alter entry paths, and move human value upward toward judgment and interaction? That is the question of the next chapter.

Chapter 7

AI and Jobs — The Big Picture

Horizon Insight: AI changes jobs by changing the tasks inside them. The real challenge is helping people move toward judgment, supervision, trust, care, and new pathways of visible experience.

In the companies I worked with, job titles often changed more slowly than tasks. A person might still be called an analyst, manager, engineer, or support specialist, but the actual work beneath the title would shift little by little. That is why I keep returning to tasks before job titles. The surface name of a job can remain familiar while the daily work underneath it is being reorganized.

The Human Question Beneath Job Change

The More Useful Question

In business, I learned that job disruption rarely arrives wearing a sign that says "disruption." It arrives first as a new feature, a faster workflow, a cheaper service expectation, or a manager asking why a process still takes three days when a competitor can do it in three hours. The job title may remain unchanged while the daily substance of the work is being hollowed out or rebuilt underneath.

That is why the better question is not simply whether AI will take a job. The better question is which parts of the job are becoming easier, cheaper, faster, or more automated, and which parts now require more human judgment than before. A person who asks that question early has time to move upward. A person who waits until the old task disappears may feel blindsided.

The question "Will AI take my job?" is understandable, but it is not the most useful place to stop. A better question is: which parts of my work are becoming easier for machines, and which parts will require more human judgment than before? That question gives a person something to do. It turns fear into diagnosis.

A young person can ask: how do I show real experience when first drafts are cheap? A middle-aged worker can ask: where can I move upward from routine execution into coordination, supervision, customer trust, and judgment? An employer can ask: how do we preserve the training ground for future leaders while using AI to remove drudgery? These questions do not remove the difficulty, but they restore agency.

In almost every audience, this question appears sooner or later. Sometimes it is asked directly. Sometimes it is hidden behind a more polite question about skills or education. But the concern is the same: will the world still have room for the work I know how to do, and will my experience still matter?

That fear deserves respect. It should not be dismissed as resistance to progress. Work is tied to income, yes, but also to pride, rhythm, usefulness, and the feeling that one has a recognized place in society.

When people hear about artificial intelligence, the first question they ask is: "Will AI take my job?" This concern is deeply understandable. For most individuals, a job is not only a source of income, but also a source of stability, identity, and social connection. Because work plays such a central role in life, any technology that threatens to change it creates a natural sense of anxiety.

To think with clarity about this issue, however, we need to move beyond immediate fear and examine how technological change has affected jobs in the past. History does not provide perfect answers, but it gives us a framework for understanding what is happening now and what may happen next.

The Waves of Technological Displacement at a Glance

Displacement Wave	Core Driver	Primary Impact Sector	Human Adaptive Guardrail
Wave 1: Industrial	Mechanical power	Physical labor and craft	Specialized skill and certification
Wave 2: Digital	Software and networks	Transactional processing	Information management and digital literacy
Wave 3: Intelligence	Generative models and agents	Cognitive synthesis and coordination	Systemic judgment and authentic voice

The Lessons of History: Creative Destruction

Economists often describe technological change using the idea of "creative destruction." This means that new technologies destroy some jobs while creating others. Over time, the overall effect has often been positive, with new opportunities emerging to replace those that disappeared.

A well-known example is agriculture. In 1900, a large portion of the workforce was employed in farming. As machines such as tractors were introduced, the need for manual farm labor declined sharply. However, this did not lead to permanent mass unemployment. Instead, workers moved into new industries such as manufacturing, healthcare, education, and services—many of which did not exist in their modern form before.

This pattern has repeated across different periods of technological change. While specific roles disappear, new categories of work tend to emerge, often requiring different skills and offering new forms of value.

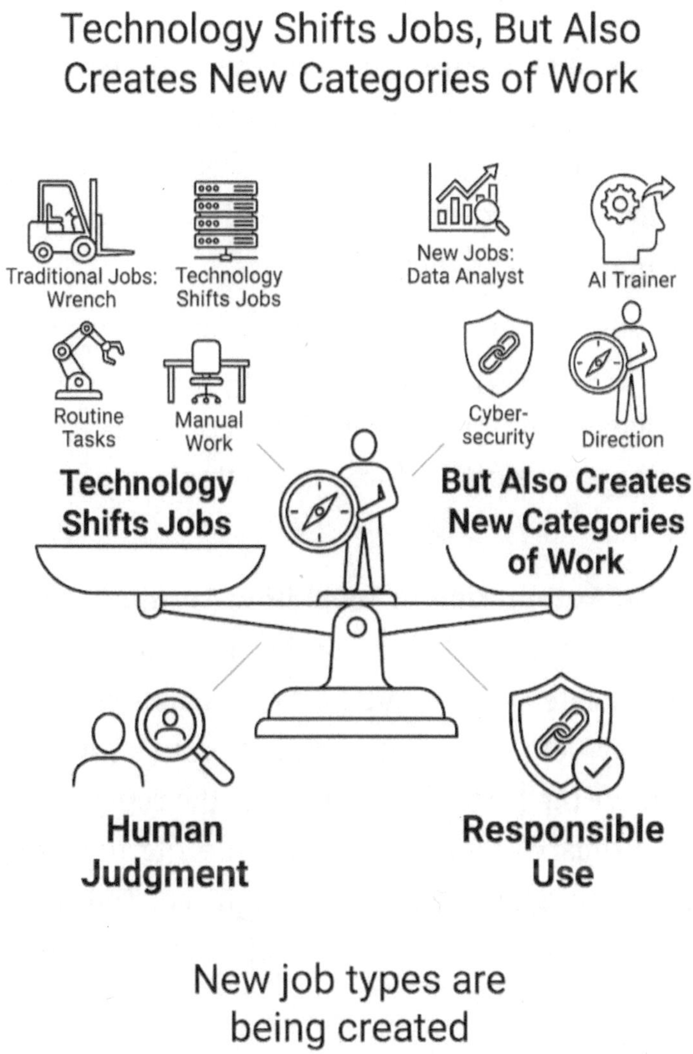

What Is Different This Time

While history offers reassurance, artificial intelligence introduces an important difference. In earlier technological waves, machines primarily replaced physical labor, and later computers replaced routine mental tasks such as calculations and record-keeping.

Artificial intelligence goes further by performing a broader range of cognitive tasks. It can write, analyze, recognize patterns, and assist in decision-making. This means it affects not only manual workers, but also professionals and knowledge workers.

Tasks such as document review, basic accounting, and certain types of writing can now be performed faster and more efficiently by AI systems. This represents a shift from automating physical effort to automating structured thinking.

From Task Automation to Workflow Automation

AI agents may deepen this change because they can automate not only isolated tasks, but parts of whole workflows. A chatbot may help write one customer email. An agent may help monitor the customer issue, draft the reply, look up the account, prepare a follow-up, schedule the next step, and alert a human when judgment is required. That is a larger change. It moves AI from task assistance toward process participation.

This matters for jobs because many roles are built around sequences of small coordinating activities. If AI agents can handle more of those sequences, then organizations may redesign work around fewer people supervising more automated processes. Some workers may become more productive and more valuable because they can manage agent-supported workflows. Others may feel pressure if their work consists mainly of routine coordination that can be standardized.

The result is not simply “AI takes jobs.” The more accurate picture is that AI changes the bundle of tasks inside jobs, and agents may change the workflow around those jobs. Human value then moves toward defining goals, handling exceptions, building relationships, making judgments, and taking responsibility for consequences.

The Rise of Human-Centric Roles

As AI takes over structured and repetitive tasks, the remaining areas of work tend to emphasize qualities that are more distinctly human. These include empathy, complex judgment, interpersonal interaction, and the ability to operate in unpredictable environments.

Roles in healthcare, education, skilled trades, and leadership may become more valuable. In addition, new roles will emerge that focus on managing and working alongside AI systems. These roles require understanding how to guide, interpret, and integrate machine-generated outputs into real-world decisions.

The important point is that work does not disappear—it changes in character. Humans move toward areas where judgment, context, and human connection matter more.

The New Work AI May Create

AI will remove some tasks and weaken some familiar job pathways, but it will also create new forms of work. These new roles may not always look like the old job titles. They will often appear around the places where machines need human direction, where automated systems need oversight, where trust must be protected, and where technology must be connected to real human needs.

Supervise machine outputs: Review automated work, examine logs, and decide when a human must step in.

Anchor domain expertise: Combine AI fluency with serious knowledge in a field that matters.

Redesign humane workflows: Make work faster without making it careless or dehumanizing.

Protect trust infrastructure: Verify identity, detect fake content, guard privacy, and preserve confidence.

Connect robotics to care: Apply physical AI to mobility, health, logistics, energy, and independence.

The new work created by AI will gather around places where machines need direction, trust must be protected, and technical power must be connected to human needs.

The important point is not that every lost job will be neatly replaced. The more realistic lesson is that human value moves toward judgment, supervision, trust, care, creativity, and the ability to connect powerful tools to real human purposes.

New work will appear, but it will not automatically reach the people who lose old work. That is why education, training, apprenticeships, career transitions, and family guidance matter. A humane society must help people move toward the new zones of value rather than simply telling them to be optimistic.

The Entry-Level Catch-22 in Practice

As Chapter 1 explained, AI may thin the first rung of professional life by compressing the tasks that once trained beginners. The practical issue here is not to repeat that warning, but to identify the response: colleges need supervised projects, internships, co-op placements, and proof-of-work portfolios; employers need junior roles that place beginners beside AI; and families need to help young people demonstrate genuine understanding rather than polished output alone.

The answer must be structural: colleges need projects, case-style work, internships, co-op placements, and proof-of-work portfolios; employers need new junior roles that put beginners beside AI rather than outside the door; and families need to help young people show real work, not only polished output.

The Catch-22 Cuts Both Ways

A company may believe it is saving money by automating junior work. In the short run, that may be true. But if too few young workers enter the pipeline, the company may eventually discover that it has weakened its own future. There will be fewer people with the lived experience, customer judgment, institutional memory, and leadership maturity needed to replace retiring workers.

This risk becomes sharper as populations age. If more experienced workers retire while fewer young workers are trained through real responsibility, the organization may face a hollowing out of its middle and senior layers. It may have advanced tools, large databases, and efficient workflows, but too few people have grown through the messy apprenticeship of real work.

For that reason, the Entry-Level Catch-22 cuts both ways. It harms young workers by blocking the doorway into experience, and it harms employers by weakening the long-term supply of experienced people. A wise society must therefore ask not only how AI can make work faster, but how young people will still become capable enough to carry institutions forward.

When AI Is Not Automatically Cheaper Than People

There is another reason the future of work may be less simple than the phrase "AI replacement" suggests: running AI is not free. A company may save a visible salary line by automating a task, but the replacement system may create other costs that are less visible at first: model usage, cloud and data-center charges, integration work, monitoring, cybersecurity, legal review, error correction, and human supervision.

This matters especially as AI moves from simple assistance toward multi-step agents. A chatbot that drafts a paragraph may be inexpensive. But an agent that handles a business process, touches customer data, uses several software systems, makes recommendations, and requires audit trails can become much more costly and fragile. The real comparison is not simply a human salary versus an AI subscription. It is a human workflow versus an AI workflow, including all hidden costs and risks.

For that reason, some companies may eventually discover that hiring and training people still makes economic sense, especially for entry-level employees. A young worker may be slower at first, but that person can learn context, build relationships, develop judgment, and become part of the future leadership pipeline. An automated workflow may perform a narrow task, but it does not grow into a manager, mentor, executive, or trusted colleague.

This does not mean AI layoffs will stop or that automation will be reversed everywhere. Some automation will be powerful and economical. But it does mean that the first wave of enthusiasm may be corrected by experience. Wise companies will learn to ask not only, "Can AI do this task?" but also, "Is AI really the best long-term way to do this work, once cost, risk, training, trust, and future talent are counted?"

The Career Chameleon

The deeper lesson is that professionals and students can no longer prepare for only one fixed job identity. In the AI age, old jobs may disappear, familiar jobs may be redefined, and new jobs may appear before schools, families, or professional associations have names for them. A person may be asked to move into work that was not part of the original degree, the original job description, or the original career plan.

That is why one of the most important human capacities may be the ability to become a career chameleon: not shallow, not rootless, and not without standards, but able to change shape as the work changes. The career chameleon carries a stable human center into shifting circumstances. That center includes adaptability, communication, relationship-building, curiosity, judgment, and the confidence to learn what is unfamiliar.

This does not mean pretending to be qualified for work one cannot responsibly do. It means being willing to enter new territory with humility and energy. A person may begin by saying, "I do not know this yet," and then learn quickly enough, ask clearly enough, build trust honestly enough, and contribute usefully enough to grow into the new role.

The false hope of the AI age is the belief that one narrow skill, one credential, or one safe category will protect a person for life. The more realistic hope is different: build a human foundation strong enough to move across changing work. The future may not reward the person who never has to change. It will more likely reward the person who can change without losing integrity, discipline, or human purpose.

No Career Category Is a Permanent Shelter

Some people now advise young people to move away from office-based careers and choose hands-on work because physical trades seem safer from AI. There is a serious truth inside that advice. Skilled trades are honorable, necessary, and often more grounded in the physical world than many routine office tasks. Society will continue to need people who can build, repair, install, maintain, and work with real materials.

But it would be a mistake to replace one oversimplified promise with another. White-collar work is not automatically doomed, and blue-collar work is not automatically immune. Robotics, sensors, autonomous machines, computer vision, and AI-supported tools will also reshape physical work over time. The safer principle is not to chase whichever job category looks protected this year. The safer principle is to help people become adaptable, capable, and deeply human across changing tools and fields.

The future does not belong simply to white-collar workers or blue-collar workers. It belongs to human beings who can keep learning, work with new tools, solve real problems, and bring judgment, character, skill, and purpose into the field where their talents can make a genuine contribution.

The Challenge of Transition

Even if new jobs emerge over time, the transition can be difficult. Changes may occur faster than individuals and institutions can adapt. People who have built careers around specific skills may find those skills less relevant in a relatively short period.

This creates both economic and psychological challenges. Workers may need to retrain, change careers, or redefine their sense of professional identity. Societies will need to support this transition through education, training, and policy adjustments.

Looking Ahead: The Risk of a Widening Gap

While history suggests that new opportunities will emerge, it does not guarantee that the transition will be smooth or evenly distributed. Some individuals will adapt quickly and benefit from new technologies, while others may struggle to keep up.

This creates the possibility of a widening gap between those who can effectively work with AI and those who cannot. If left unaddressed, this gap could lead to broader social and economic divisions.

This widening gap leads directly to the next question: whether the gains of AI can be shared broadly enough to preserve dignity and social trust.

Chapter 8

Inequality in the Age of AI

Horizon Insight: AI amplifies starting position. A humane society must build bridges, so the technology does not create a two-speed future divided between those who direct AI and those who merely live under it.

This is also where my experience with enterprise customers shapes my concern. The first organizations to benefit from powerful technology are often those that already have capital, data, management discipline, and technical staff. Smaller organizations and ordinary workers may benefit too, but not automatically. Without deliberate effort, AI can widen differences that were already present.

When Technological Power Magnifies Existing Differences

Inequality can sound like a policy word, but it is often lived in very concrete ways. One student has parents who help her use AI as a tutor; another uses it only as a shortcut. One worker has time and support to learn new tools; another is already exhausted by two jobs and family obligations. One company has data, capital, and technical staff; another is still trying to keep up with ordinary operations.

These differences matter because AI does not arrive in a neutral world. It arrives inside the lives people already have.

Whenever a powerful new technology appears, people often focus first on what it can do. But another question is just as important: who is positioned to benefit most from it? Artificial intelligence is not arriving in an equal society. It is arriving in a world already marked by differences in wealth, education, access, infrastructure, confidence, social networks, institutional support, and political power.

That matters because technology rarely lands on a level playing field. It tends to magnify what is already there. Those who begin with stronger tools, stronger skills, stronger networks, and stronger positions can often turn a new technology into even greater advantage. Those who begin with less may gain some benefits, but not in the same way and not at the same scale.

That is why inequality belongs near the center of any serious discussion of AI. The question is not only whether artificial intelligence will make society more productive. It is whether the gains will be shared broadly enough to create wider dignity and participation — or concentrated narrowly enough to deepen social division.

AI as an Amplifier

One of the clearest ways to understand AI is to see it as an amplifier. It amplifies capability. It amplifies speed. It amplifies leverage. It amplifies the output of the person, team, or institution using it well. But amplification is neutral with respect to justice. It does not ask whether the starting point was fair.

If a highly educated professional has access to strong models, good data, organizational support, and the freedom to experiment, AI may make that person dramatically more powerful. If a company already has capital, infrastructure, brand, talent, and customer data, AI may allow it to expand those advantages even further. If a nation already controls chips, compute, energy, cloud infrastructure, and research ecosystems, AI may deepen that lead.

That is why the central inequality question is not simply who uses AI and who does not. A more important question may be: who directs AI, who owns AI, who builds on top of AI, and who merely lives under systems shaped by AI but controlled by others?

The same tool can accelerate people differently when the starting lines are unequal.

The reader should see that AI does not arrive on neutral ground; it can widen old gaps unless people build bridges across them.

The Difference Between Using Tools and Controlling Systems

Many people will use AI tools. They may become more efficient, more creative, and more informed. That matters. But using a tool is not the same as controlling a system. A person who uses AI to write faster gains convenience and perhaps some leverage. A person or institution that owns the models, the compute, the customer relationship, the data pipelines, and the distribution channels stands at a very different level of power.

This distinction matters because it helps explain why AI may widen inequality even if millions of people have access to AI. Broad access by itself does not fix the imbalance. Many people may gain modest benefits while a smaller number gain extraordinary leverage.

That is why the AI age may sharpen the old distinction between participation and control. One can participate in a technology-driven system without meaningfully shaping its direction. And those who shape direction usually capture the greatest share of value.

Position Matters More Than Ever

In ordinary conversation, people often speak of inequality mainly in terms of income. But income is only one surface measure. A deeper concept is position. Position includes access to education, ability to adapt, freedom to experiment, institutional support, financial cushion, network quality, geographic location, and proximity to centers of innovation and decision-making.

Artificial intelligence increases the importance of position because the transition rewards not only hard work, but strategic placement. A person in a strong position can learn faster, test tools sooner, absorb mistakes more safely, and connect their new capabilities to real opportunities. Someone in a weaker position may face the same technology from a place of pressure rather than possibility.

That is one reason the AI age may feel unfair even when the tools are technically available to many people. Availability is not the same as readiness. Access is not the same as position. And position may increasingly determine who converts AI from a curiosity into a life-changing source of leverage.

Education, Confidence, and the Adaptation Gap

This is especially important for older adults and for people who did not grow up inside digital culture. A person may be intelligent, experienced, and wise, yet still feel hesitant in front of a new tool. Without encouragement and patient guidance, hesitation can become exclusion. With support, that same person may discover that AI can extend independence, learning, and usefulness.

Another source of inequality lies in adaptation itself. Artificial intelligence rewards not only technical knowledge, but also confidence, curiosity, and the willingness to experiment. People who are comfortable testing new tools, revising workflows, and learning continuously are likely to benefit more than people who feel intimidated, overworked, or excluded from technical culture.

This creates an adaptation gap. A strong professional with time and freedom may integrate AI into work quickly. A teacher with too many responsibilities, a worker in a rigid job, or an older adult with less exposure to new digital systems may move more slowly, even if they are equally intelligent and equally capable of learning under better conditions.

That is why inequality in the AI era is not just about raw intelligence. It is about whether society gives people fair chances to adapt. If the burden of adaptation falls entirely on individuals while institutions move ahead without support, then inequality will deepen not because some people are less worthy, but because the structure of transition itself is unequal.

Capital, Ownership, and the New Advantage

Artificial intelligence also has a capital dimension that should not be underestimated. Advanced AI depends on chips, data centers, cloud systems, specialized talent, and enormous energy resources. These are not distributed equally. They require large investments and often concentrate in a limited number of firms and countries.

That means the people and institutions closest to ownership may capture the largest gains. If AI makes workers more productive but ownership remains narrow, then rising productivity may not translate into broad prosperity. Instead, it may translate into rising returns for those who already control the infrastructure of the next economy.

That is why the discussion of inequality cannot stop at education and adaptation. Ownership matters. Capital matters. Control over infrastructure matters. A society that ignores these questions may discover too late that it has allowed extraordinary productive power to become concentrated in too few hands.

The Risk of a Two-Speed Society

If these patterns continue, society may begin to split into two broad speeds. One group moves quickly with the technology. They gain leverage, confidence, and economic advantage. They understand how to work with AI, how to direct it, and how to use it to multiply their reach. The other group moves more slowly. They may use AI superficially, fear it, resent it, or remain structurally distant from the ecosystems where the main gains are being captured.

A two-speed society is dangerous not only because of the income gap it may create, but because of the psychological and political consequences that follow. People who feel excluded from the future often do not interpret the situation in technical terms. They experience it as humiliation, invisibility, and loss of standing. When enough people feel that way, legitimacy weakens.

That is why inequality is not merely an unfortunate side effect to be discussed later. It is one of the core questions of the AI age because it affects whether a society remains stable, coherent, and morally defensible while adopting extraordinary new tools.

Beyond Individuals: The Global Divide

The problem is not limited to individuals within one country. AI may also widen the divide between nations. Countries that control advanced semiconductor supply chains, cloud infrastructure, frontier research, robotics ecosystems, and reliable energy may gain extraordinary strategic advantage. Countries that lack these foundations may become increasingly dependent on external systems they do not control.

This creates the possibility of an AI-rich world and an AI-poor world. Wealthier countries may gain faster productivity, stronger military systems, more advanced healthcare, better educational tools, and greater bargaining power in global markets. Poorer countries may remain consumers of intelligence systems built elsewhere, subject to rules, pricing, and dependencies shaped by others.

That possibility should matter morally as well as strategically. If artificial intelligence becomes one of the great engines of 21st-century power, then the distribution of AI capability may help determine not only business outcomes, but the future structure of global inequality.

Why This Is a Moral and Political Question

At some point, inequality stops being only an economic statistic and becomes a moral question. What kind of society are we willing to build? One in which powerful tools mostly enlarge the already strong? Or one in which those tools are used to widen real opportunity, dignity, and participation?

This question also becomes political because large disparities in power and security eventually shape trust in institutions. If people come to believe that AI is mainly a system for concentrating advantage, then resistance will grow. Some of that resistance will be wise. Some of it may be reactive. But either way, legitimacy will be at stake.

That is why the age of AI requires more than innovation. It requires social imagination and moral seriousness. A civilization cannot rely only on technical brilliance if it neglects the human terms on which that brilliance will be lived.

Artificial intelligence may become one of the great amplifiers of inequality in modern history unless societies respond deliberately. It magnifies capability, but it also magnifies starting position, ownership, leverage, and control. That is why the question is not only whether AI creates value. It is who shares in that value and under what terms.

This leads directly to the next chapter. If AI is built partly on collective human inheritance and increasingly helps generate extraordinary productivity, then should the gains belong only to those who control the systems? Or should part of that gain flow back to society more broadly in the form of dignity, security, and participation? That is the argument of the AI Dividend and the possibility of a new social contract.

Chapter 9

The AI Dividend and a New Social Contract

Horizon Insight: If AI is built partly on shared human inheritance and creates enormous value, society must ask how some of that value can support dignity, participation, and a floor of freedom.

I do not think about the AI dividend as a slogan. I think about it as a practical question raised by a productive society: if machines and systems help create enormous new value, how can ordinary people continue to live with dignity, purpose, and purchasing power? A society cannot be healthy if productivity rises while too many people feel unnecessary.

When Productivity and Dignity Begin to Separate

If artificial intelligence helps produce more economic value with less direct human labor, then society eventually faces a deeper question than efficiency: who shares in that value? For a long time, modern economies were organized around a powerful assumption: if you work, you earn. That assumption never described reality perfectly, but it remained central to how people understood fairness, contribution, and self-respect.

Artificial intelligence may weaken that link. If large amounts of value can increasingly be generated by systems owned by relatively few firms and guided by relatively few people, then the old connection between labor and livelihood may no longer be strong enough to support a stable and humane social order.

That is why the idea of an AI Dividend matters. It is not merely a policy concept. It is a dignity concept. It asks whether a civilization transformed by intelligent systems can preserve broad participation, self-respect, and freedom even if the older structure of work begins to change.

Why the Old Model May No Longer Be Enough

In the industrial world, the social contract rested heavily on employment. People contributed labor, and in return they received wages, status, routine, and a place in the productive structure of society. The system had many injustices, but it still provided a recognizable framework. Work linked people to income and, through income, to participation.

Artificial intelligence may disrupt that framework more deeply than earlier tools because it can increase output without always requiring proportionately more human labor. A smaller number of people, supported by powerful systems, may be able to generate very large economic value. If that happens, society cannot simply tell everyone else to work harder and pretend the structure remains unchanged.

This does not mean work disappears. It means that the old model may no longer be enough by itself. If the productive engine of the economy changes, then the distribution system that supports legitimacy may need to change as well.

What an AI Dividend Is — and Is Not

The phrase AI Dividend should be understood carefully. It does not mean charity. It does not mean a reward for passivity. And it should not be confused with the idea that people no longer need purpose, contribution, or effort. The deeper idea is that if AI is built partly on collective human inheritance — centuries of knowledge, language, institutions, public research, shared culture, and social stability — and if it increasingly helps generate extraordinary productivity, then part of that productivity may reasonably be understood as having a broader social claim.

Seen this way, an AI Dividend is closer to the logic of shared inheritance than to the logic of welfare. Welfare is often framed culturally as help for those who cannot make it on their own. A dividend is framed differently. It implies that one has a rightful share in a system that produced value with the help of common foundations.

That distinction matters because language shapes dignity. A society that distributes support in a way that feels humiliating will deepen resentment. A society that distributes support in a way that preserves standing may help protect social cohesion.

The Floor of Freedom

What Freedom Should Make Possible

A floor of freedom should not be imagined as a hammock in which a society goes to sleep. It should be imagined as a stable platform from which people can stand up again. A person with some security can retrain without terror, care for an elderly parent without financial collapse, help raise grandchildren, start a small company, serve in a community organization, recover from illness, or move from a declining occupation into a new field.

This is why the phrase matters. Freedom is not only the absence of hunger or panic. It is the practical room to contribute. When people are crushed by insecurity, they often cannot think beyond survival. When they have enough stability, they can take the kinds of risks and responsibilities that make a society more creative, more caring, and more resilient.

The floor of freedom is not meant to create a society of spectators. It is meant to give people enough stability to contribute in ways that an anxious survival economy often prevents. A person with a floor beneath them may retrain for a changing field, start a small company, care for an aging parent, raise children with more patience, volunteer in a community, mentor young people, return to school, build a local service organization, or take the risk of creating something useful that does not yet have a guaranteed salary attached to it.

This is why the language of freedom matters. A floor of freedom should not be imagined as a couch. It should be imagined as a foundation. The point is not to remove responsibility. The point is to make responsibility livable. A society that uses AI abundance only to enrich the already secure may become more productive and more brittle. A society that uses some of that abundance to widen participation may become both more humane and more inventive.

Freedom here does not mean luxury. It means the ability to think, choose, care for others, learn, retrain, create, and participate without being crushed by constant economic panic. A society that gives its people such a floor does not remove the need for discipline or contribution. It removes some of the humiliation and terror that make human development harder.

A floor of freedom should give people room to contribute, not permission to disappear. It can help someone retrain for new work, start a small company, care for children, support aging parents, serve a neighborhood, volunteer in a community, recover from disruption, or take the first step toward a more meaningful form of contribution.

This is one reason the AI Dividend should be understood morally as well as economically. It is not only about money. It is about whether a civilization uses new productivity to widen human freedom or only to deepen the advantage of those already strongest.

Support Without Passivity

A good social floor steadies people so they can keep contributing.

Society can only thrive when basic human dignity is preserved, and a floor of freedom is a key structure to support that.

Why This Is Not a Case Against Work

Some people worry that any dividend-like system would weaken work ethic or encourage passivity. That concern should be taken seriously. Human beings do need challenge, effort, contribution, and responsibility. A civilization that merely distributes consumption without encouraging purpose would create a thin and dangerous kind of life.

But that is not the argument here. The argument is not that work stops mattering. The argument is that if the structure of work changes, then the structure of dignity must be protected in another way. A person should not have to prove their worth only by fitting into an older labor model that technology itself is beginning to transform.

In other words, the AI Dividend is not meant to replace the search for meaning. It is meant to prevent technological abundance from turning into mass insecurity while society searches for a better way to organize contribution, freedom, and belonging.

Possible Forms of Distribution

The exact form of an AI Dividend could vary. It might involve direct cash payments, public ownership stakes in certain AI-related infrastructures, sovereign-style technology funds, stronger public services, educational credits, healthcare support, or combinations of these. Different societies may design different models depending on culture, law, and political structure.

The purpose of this book is not to lock the reader into one specific blueprint. The deeper point is that the question of distribution can no longer be treated as secondary. If AI produces enormous gains, a society will have to decide whether those gains remain narrowly concentrated or whether some significant portion flows back to the broader public.

That is why the policy question cannot be separated from the civilizational question. The design details matter, but the principle comes first: when a technological system built partly on shared human inheritance creates vast new wealth, broad human participation in that wealth becomes a question of legitimacy.

The Difference Between Welfare and Participation

The moral danger in many modern discussions of redistribution is that support is framed in a way that separates recipients from full participation in society. They are helped, but not honored. They receive aid, but not standing. The result is often stigma, resentment, and a sense that the social contract has weakened.

That is why the AI Dividend must be understood not as passive rescue, but as participation in a transformed economy. A dividend says: this society created extraordinary productive power partly through common inheritance, and its members therefore share in the gains. That framing is far more compatible with dignity than a model that treats large parts of the population as tolerated dependents.

If society gets this wrong, it will not only create economic stress. It will create a culture of humiliation. And that can become politically explosive.

A New Social Contract

At the human level, this chapter is really about the possibility of a new social contract. A social contract is the set of implicit agreements through which people accept the legitimacy of a system. They accept laws, institutions, taxation, and hierarchy because they believe, at least in broad terms, that the system remains livable and fair enough.

If artificial intelligence changes how value is created, then the social contract may need renewal. The question is no longer only how to encourage innovation. It is also how to preserve legitimacy in a world where innovation may dramatically reduce the economic centrality of some forms of labor.

A new social contract would not abolish effort, excellence, or ambition. It would preserve them. But it would also recognize that a technologically transformed society cannot remain stable if growing numbers of people feel structurally unnecessary while abundance rises around them.

The Hope and the Danger

The hopeful possibility is clear. If AI creates extraordinary wealth, humanity may have a chance to reduce certain forms of desperation, widen freedom, support education and caregiving, and allow more people to build lives shaped not only by survival but by contribution and meaning.

The danger is equally clear. The gains may remain concentrated. The winners may insist that all value belongs only to those who controlled the systems. Broad insecurity may then rise at the same time as visible abundance, producing resentment, instability, and a crisis of legitimacy.

The difference between those two futures will not be decided by technology alone. It will be decided by political courage, institutional imagination, and the moral seriousness with which societies confront the distribution question.

The economic chapters of this book have moved through a deliberate sequence. We began with the changing nature of the economy, then looked at jobs, then inequality, and now the question of a dividend and a new social contract. Taken together, they suggest that the AI age is not only about growth. It is about participation, dignity, and legitimacy.

Part II Summary

The Economic Shock

Part II showed that the AI economy is not only a productivity story. It is a question of jobs, entry pathways, inequality, bargaining power, and whether technological gains will support broad dignity.

Part III Introduction

The Core of Human Meaning and Response

This part gathers the heart of the book: the human being as compass, the skills and education needed for judgment, and the deeper question of meaning when work and production are transformed.

Chapter 10

The Role of Humans in the Age of AI

> ***Horizon Insight: The machine may become the engine, but the human being must remain the compass. Power without direction is not progress.***

This chapter is the philosophical hinge of the book. It does not yet offer a training plan or an education reform agenda. It answers the deeper question first: when the machine gains more power, what must human beings refuse to surrender?

Returning to the Central Question

The Boardroom, the Classroom, and the Concert Hall

The same lesson appeared in three very different places in my life. In the boardroom, the hardest problems were rarely solved by software alone; they required trust, patience, translation between groups, and judgment about what the customer truly needed. In the classroom, information was never enough; students needed formation, encouragement, and the discipline to explain ideas clearly. In the concert hall, technique mattered greatly, but technique without human meaning never fully reached the listener.

These experiences taught me that human value often appears at the point where capability must be joined to purpose. AI may become a powerful engine in all three worlds. But the engine still needs a human compass: the executive who understands the people behind the process, the teacher who sees the student behind the assignment, and the singer who knows that a phrase must carry more than sound.

After all the discussion of technology, economy, jobs, inequality, and meaning, we return to the question that quietly stands behind every chapter: what should human beings now become?

I often think about this through music. A singer can have perfect notes and still give a lifeless performance. What makes the song human is breath, phrasing, restraint, courage, and the willingness to carry meaning through the voice. In the same way, a society can become very efficient and still become spiritually thin if it forgets why efficiency matters.

AI can map acoustic frequency, imitate tone, and reproduce the surface pattern of a musical line. But it cannot feel the vulnerability of controlling breath before a live audience, or carry grief, gratitude, and triumph through a human body into a room of listeners. That is the difference between replication and presence.

The human role in the age of AI is therefore not only to check the machine. It is to keep the music human.

From the beginning of this book, one question has stood at the center: in a world where machines can do more and more, what is the role of humans? The earlier chapters approached that question indirectly by examining artificial intelligence itself, the economy, work, inequality, meaning, and the dangers of a thinner kind of life. This chapter now addresses the question directly.

The importance of asking it plainly is hard to overstate. Whenever a powerful technology arrives, the first instinct is often to focus on the machine: what it can do, how fast it improves, how far it might go, and what it may replace. But if we stop there, we ask the wrong question. The central issue is not only what machines can do. The central issue is what human beings should do in a world shaped by increasingly capable machines.

That shift in emphasis matters because it restores agency. Technology may change the conditions of life, but it does not answer the question of what life is for. That answer remains human.

The Machine as Engine, the Human as Compass

One of the clearest ways to think about the human role in the AI age is this: the machine may increasingly function as an engine, but the human being must remain the compass.

An engine provides power. It helps move things. It accelerates execution. It can take a goal and help generate routes, options, analyses, drafts, outputs, and increasingly even multi-step plans. But an engine does not decide where one ought to go. It does not decide what is worth pursuing. It does not decide which tradeoffs are morally acceptable. And it does not decide what should be protected even when efficiency argues otherwise.

That is the work of a compass. A compass gives direction. It does not provide raw power, but it makes power meaningful. In human terms, the compass involves judgment, moral orientation, wisdom, responsibility, and the ability to choose ends rather than merely accelerate means.

Over the years, in AI, teaching, and business, I have come back to this lesson repeatedly: tools become more powerful much faster than human beings become wise. That is why direction matters so much.

That is why the human role does not disappear as the machine becomes stronger. It becomes more important. The more powerful the engine, the more dangerous it is without a compass.

Power Needs a Compass

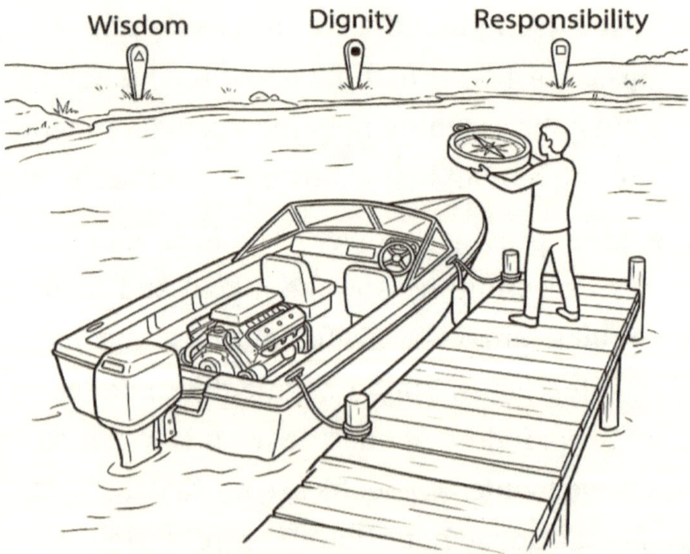

Capability creates motion;
human judgment gives it direction.

The reader should see the central human responsibility: powerful tools need human direction, not passive admiration.

Why Judgment Rises in Importance

As AI takes over more routine execution, the relative importance of judgment rises. Judgment is not simply intelligence in the abstract. It is the disciplined ability to interpret context, weigh consequences, distinguish the important from the trivial, recognize what cannot be reduced to a rule, and make decisions under conditions of uncertainty.

This matters because real life is not made of standardized prompts alone. Real life involves ambiguity, conflicting goods, emotional complexity, hidden incentives, and moral consequences. A system can provide analysis, but it cannot fully answer for the human meaning of the choice.

This is one reason highly capable AI does not eliminate the need for human beings. It increases the premium on those human capacities that cannot be reduced to pattern generation alone. Where the world becomes more automated, the need for interpretation often becomes greater, not smaller.

Responsibility Cannot Be Outsourced

Closely tied to judgment is responsibility. A machine may produce a recommendation, but it does not bear the moral burden of what follows. If a doctor uses AI support in diagnosis, the responsibility still rests with the doctor. If a judge, teacher, manager, parent, or political leader uses intelligent systems, the moral accountability does not migrate into the machine. It remains with the person.

This matters because one of the temptations of the AI age will be to hide behind technological outputs. A human being may begin to say, explicitly or implicitly, "The system recommended it." But the presence of a recommendation does not erase the burden of human choice.

In fact, the more intelligent the system appears, the more disciplined the human must become about not surrendering conscience to convenience. A civilization that confuses assistance with absolution will become morally weaker even as its tools become more powerful.

As AI agents become more capable, this responsibility becomes even more concrete. The human role is not only to use such systems, but to supervise them. A person must learn how to set direction, define limits, observe behavior, catch errors, and remain accountable for outcomes. The agent may carry out steps, but the human being must still provide the compass.

This means that the modern worker, manager, teacher, doctor, lawyer, parent, or citizen cannot treat an AI agent as a magical servant. The agent must be watched, questioned, and corrected. It may behave well most of the time and then fail in a subtle but consequential way: acting too broadly, missing context, using the wrong information, or presenting an answer with more confidence than it deserves. Human responsibility therefore includes learning how the agent behaves, where it is strong, where it is weak, and where it must not be allowed to act without review.

The Human Role Moves Upward

In many fields, the human role will not vanish. It will move upward. As the lower layers of drafting, searching, sorting, formatting, comparing, coding assistance, and routine analysis are increasingly supported by AI, human beings will be judged more by how well they define goals, evaluate results, manage ambiguity, build trust, communicate with other humans, and take responsibility for consequences.

This upward movement will not always feel comfortable. Many people built careers within older layers of work that are now being compressed. The transition may be painful. But the movement itself is important to understand. AI does not only replace. It also reveals which human capacities were being hidden beneath routine tasks.

That is one reason the age of AI may be both disruptive and clarifying. It may force us to see more honestly what kinds of human strengths still matter most.

What the Machine Cannot Fully Be

Artificial intelligence may become astonishingly capable in many domains. It may imitate conversation, generate art-like outputs, solve structured problems, and assist in decisions that once demanded substantial human effort. But even then, there remain features of human existence that cannot be captured fully by fluent output alone.

A machine does not stand in mortality. It does not love. It does not suffer. It does not bear guilt. It does not keep faith. It does not age beside another person. It does not know the inward texture of grief, sacrifice, responsibility, loyalty, repentance, or reverence. It can describe such things. It can pattern-match them. But description is not the same as lived reality.

This matters because some of the most important human roles arise precisely in those regions of life where information alone is not enough: raising children, accompanying the vulnerable, leading under moral strain, reconciling conflict, preserving trust, transmitting wisdom, and embodying courage in the presence of fear.

The Return of Human Depth

Paradoxically, the rise of AI may make human depth more visible rather than less. For many years, modern life often rewarded speed, efficiency, visible productivity, and the external markers of competence. Those things still matter. But as machines become more capable in those dimensions, qualities that once seemed secondary may become more central.

Depth becomes more visible when fluency can be imitated. Character becomes more visible when outputs can be generated. Presence becomes more visible when conversation can be simulated. Wisdom becomes more visible when information is abundant.

Seen this way, the AI age may become not only a test of human utility, but also a test of human substance. What remains when the machine can do much of what once made people seem impressive? This question may feel threatening, but it may also be the beginning of a healthier civilization if it drives us back toward what is genuinely worth cultivating.

The Human Task Is Not to Compete Like a Machine

One of the great mistakes people make in times of technological change is to try to compete with the machine-on-machine terms. They become more frantic, more reactive, more output-driven, more willing to reduce themselves to speed, volume, or constant availability. That path usually leads to exhaustion and imitation.

The better path is to rise toward what is most deeply human. Instead of trying to be a cheaper algorithm, a person must become more clearly and fully a person: more able to judge, more able to care, more able to communicate, more able to integrate knowledge into wisdom, more able to bear responsibility, and more able to orient life toward ends that are actually worthy.

This is not romantic language. It is practical strategy. In the long run, societies that cultivate deep human capacities will use powerful systems more wisely than societies that merely produce people who know how to press the buttons.

What Must Be Strengthened

If the human role is to function as compass rather than engine, then the deepest need is not one more technical trick. The deepest need is human formation: steadier judgment, clearer conscience, longer attention, moral courage, truthful speech, and the ability to remain responsible when convenience invites surrender.

The following chapters translate that human center into more practical terms. First come the skills that make a person useful and trustworthy in AI-shaped work. Then come the educational changes needed to form such people. Only after that can the book turn to advice for each generation.

The role of humans in the age of AI is not to disappear, not merely to survive, and not to imitate the machine. It is to provide what the machine cannot supply on its own: direction, judgment, conscience, responsibility, and forms of presence rooted in lived human reality.

Chapter 11

What Skills Will Matter in the Age of AI

Horizon Insight: AI fluency is only the beginning. The durable human advantages are judgment, communication, trust, adaptability, ethical reasoning, domain depth, and character.

This chapter is deliberately practical. The previous chapter named the human compass; this chapter asks what that compass looks like in daily work, school, family, and civic life. These are the capabilities a person can practice, observe, and strengthen.

As a professor, I saw students who were technically bright but sometimes too narrow. As an employer, I saw people rise because they could explain, listen, persuade, and earn trust. As a singer, I learned that technique matters, but technique without expression does not move people. These experiences all point toward the same conclusion: in the age of AI, the most valuable human skills will often be the ones that connect knowledge to judgment, communication, and responsibility.

When Capability Becomes Cheap, What Becomes Valuable?

Skills as a Human Portfolio

I like the word portfolio here because the AI age will punish narrowness. A person may need technical fluency, but also the ability to write, speak, listen, verify, negotiate, learn, and work across disciplines. No single skill will be enough by itself. The durable person will carry a balanced set of human strengths that can be recombined as work changes.

In my own career, the people who advanced were rarely those who possessed only one isolated competence. They could connect the technical question to the customer question, the customer question to the business question, and the business question to the human question. That connective intelligence may become even more important when AI supplies more of the first draft.

The future will not reward only the person with one isolated skill. It will reward people who carry a portfolio of human strengths: one area of real depth, enough breadth to connect that depth to the world, enough communication to make others understand it, enough judgment to use tools well, and enough character to be trusted with power. This is not a fashionable skill list. It is a way of remaining useful when tasks keep shifting.

I have seen this pattern repeatedly. The person who rises is often not the narrow specialist who cannot explain the work, and not the broad talker who has no depth. It is the person who can connect depth to people, purpose, and consequence. AI makes that combination more important because it can produce surface fluency. Human beings must bring grounded understanding.

This question is practical, but it is also deeply personal. Many readers are not asking only what skill to learn next. They are asking what part of themselves will still matter when machines become so capable.

If artificial intelligence can increasingly draft, summarize, compare, generate, and assist in a wide range of cognitive tasks, then many people naturally ask what skills will still matter. This is the right question, but it is often answered too narrowly. The future will not belong only to those who learn a few technical tricks. It will belong more often to those who can combine intelligent tools with distinctly human depth.

That means the skill question cannot be reduced to software familiarity alone. AI fluency matters, but it is not enough. In the years ahead, the most valuable people will often be those who know how to use powerful systems without surrendering judgment, purpose, responsibility, and human presence.

This chapter asks a practical question with a broader meaning: what capacities become more important, not less, as machine capability rises?

AI Fluency Matters — But It Is a Beginning, Not an End

One practical answer is that AI fluency matters. A person who refuses to understand these tools at all will often be at a disadvantage. They may work more slowly, miss opportunities, fail to see new possibilities, and struggle to participate in environments where AI-assisted work becomes normal.

AI fluency means more than opening an application and typing a prompt. It means understanding what these systems are good at, where they are weak, how to guide them productively, how to check their outputs, and how to incorporate them into a workflow without becoming dependent on them.

But AI fluency is only a starting point. If everyone eventually gains some access to intelligent tools, then the more decisive differences will lie elsewhere. The real question becomes what a person can add that the tool cannot supply by itself.

Part of AI fluency is understanding the basic way these systems work. A person does not need to know the mathematics of machine learning, but they should know that AI learns from patterns, generates likely outputs, and can be fluent without being fully reliable. That simple understanding changes how a person uses the tool. Instead of being dazzled by polished language, the user learns to ask: What is the source? What might be missing? What assumptions are hidden? Where does human judgment still have to enter?

This is especially important for parents, grandparents, teachers, professionals, and citizens who may feel that AI is too technical for them. They do not need to become programmers. But they do need enough literacy to supervise the tool, protect themselves from error or manipulation, and remain part of the family and civic conversation. In the AI age, basic AI literacy becomes like basic financial literacy or media literacy: not everyone is a specialist, but everyone needs enough understanding to avoid being misled.

The Human Capabilities Employers Will Notice

As AI fluency becomes more common, employers and institutions will not stop at the question, “Can this person use AI?” Many people will be able to use a chatbot, draft with a model, or ask a tool for a summary. The more important question will be whether the person can keep creating value as the work itself changes.

That means the durable human advantage will include self-knowledge, adaptability, resilience, curiosity, and the ability to receive correction without becoming defensive. It will also include systems awareness: the ability to see how one task connects to customers, colleagues, costs, risk, quality, trust, and the larger purpose of the organization.

The most valuable people will not merely defend yesterday's job description. They will notice where old workflows are breaking, where AI can remove drudgery, where human judgment must remain central, and where they can move upward into higher-value contribution before someone else has to tell them. Seen this way, the AI age rewards not only technical skill, but also maturity, imagination, humility, and the courage to keep learning.

Do Not Measure Machine Activity as Human Progress

There is another warning here. In the AI age, institutions may be tempted to measure what is easiest to count rather than what truly matters. A company may count how often employees use AI. A school may count how quickly students produce assignments. A platform may count engagement. But more machine activity does not automatically mean better judgment, deeper learning, stronger trust, or more useful work.

The human question is not simply whether AI was used. The better question is whether AI strengthened human capability. Did it improve accuracy, responsibility, learning, service, creativity, or wisdom? If not, the activity may be noise rather than progress.

Managing AI Agents

As AI moves from answering questions to acting as an agent, another practical skill becomes important: knowing how to manage AI agents. This is not the same as merely writing a clever prompt. It means defining goals clearly, setting boundaries, deciding what information the agent may use, deciding what actions require permission, and reviewing the work before consequences become real.

A person who manages AI agents well will not simply ask the system to "handle it." They will specify the desired outcome, provide context, set limits, require checkpoints, ask for explanations, and verify the final result. In a workplace, that may mean supervising several agent-supported workflows. In personal life, it may mean letting AI help organize travel, documents, appointments, or research while keeping private information, money, and important decisions under human control.

For that reason the human role moves upward. The person becomes less only an executor of steps and more a designer of goals, boundaries, review points, and accountability. In the age of agents, one of the most valuable skills may be the ability to delegate to machines without surrendering judgment to them.

Judgment

Judgment is one of the most important skills of the AI age. It includes the ability to ask the right question, define the real problem, distinguish signal from noise, identify what matters deeply, and recognize when an answer that sounds plausible is actually weak or misleading.

AI may generate options quickly. Judgment determines which option is worth taking seriously. AI may summarize large amounts of information. Judgment determines what is missing, what is distorted, and what consequences follow from acting on the summary.

In a world of increasingly abundant outputs, the person who can judge well becomes more valuable, not less. That is because abundance creates a new problem: not lack of information, but overload, confusion, and persuasive error.

Communication

Communication will also matter more. Some people assume that because AI can generate fluent language, human communication will matter less. The opposite may often be true. As machine-generated language becomes more common, the human ability to communicate with clarity, sincerity, emotional intelligence, and persuasive presence becomes more distinctive.

Good communication is not only about words on a page. It includes listening, tone, timing, framing, explanation, empathy, and the ability to speak into the actual condition of another human being. AI may help prepare language, but it does not fully replace the relational depth that gives communication trust.

This means people who can communicate well across differences, calm confusion, align teams, explain difficult ideas, and speak credibly in moments of uncertainty may become more rather than less valuable.

Trust and Human Presence

As more work becomes mediated by systems, trust becomes even more important. People want to know not only whether a result is technically competent, but whether the person behind it is reliable, careful, honest, and accountable.

That is why human presence matters. Presence includes the ability to reassure another person, stay with a problem, remain calm under pressure, notice emotional reality, and convey that one is actually there rather than merely generating a surface performance.

In caregiving, teaching, leadership, negotiation, counseling, medicine, and many other fields, this human presence will remain one of the clearest differences between assistance and relationship. The AI age may make that difference more visible.

I saw this repeatedly in my own company. Some employees were technically capable, but those who could explain ideas clearly, listen to customers, calm tension, present with confidence, and earn trust usually went much farther. The same pattern appeared when many Fortune 500 companies purchased products from my company for optimization, communication, and migration to the enterprise cloud. The best relationships were not built on technical answers alone. They were built by people who could understand the customer's real concern, communicate across functions, and make others feel that the problem was being handled responsibly. In the era of AI, those human skills become even more critical, not less, because machines can help with tasks but human beings still create trust.

Adaptability and Learning

Another skill that rises in importance is adaptability. The old model of learning a profession once and then repeating the same methods for decades becomes less reliable when tools, expectations, and workflows keep shifting. People will need to learn continuously, not because endless novelty is fashionable, but because flexibility has become part of survival.

Adaptability does not mean rootlessness. It means being able to absorb new tools, rethink old assumptions, and move from one form of competence to another without collapsing internally. It requires humility, curiosity, and the willingness to revise one's methods without losing one's center.

This is one reason lifelong learning becomes so important. The adaptable person is not only more employable. They are also less likely to become psychologically brittle when the world changes.

Breadth and Domain Depth

The future will likely reward people who combine breadth with domain depth. Domain depth still matters because real expertise cannot be replaced by surface fluency alone. A doctor still needs medicine. A teacher still needs real understanding of students and subject matter. A lawyer still needs legal reasoning. An engineer still needs engineering judgment.

But breadth matters too, because the age of AI rewards people who can connect disciplines, translate between technical and human domains, and place narrow outputs into a larger context. The strongest people may not be the narrowest specialists or the shallowest generalists. They may be those who have one serious area of strength while remaining broad enough to think across fields.

This is one reason the major plus one idea is so useful. A person needs real depth in something, but also enough adjacent reach to work powerfully with new tools and changing environments.

Ethical Reasoning

Ethical reasoning becomes more important as systems become more powerful. The question is no longer only whether something can be done. It is whether it should be done, on what terms, at whose expense, and with what safeguards.

As AI enters education, hiring, medicine, law, surveillance, finance, warfare, and public life, people who can think seriously about fairness, harm, dignity, responsibility, and long-term consequences will become indispensable. A society full of technically skilled people who lack ethical seriousness will not handle powerful systems wisely.

Ethical reasoning is therefore not a luxury skill. It is part of the practical competence required by the AI age.

Skills That Steer the Machine

Durable skills guide, interpret, and humanize powerful tools.

The most durable skills are those that guide, interpret, and humanize intelligent tools.

Resisting Manipulation

Another increasingly important capacity is the ability to resist manipulation. Artificial intelligence can already help generate persuasive language, personalized messaging, synthetic media, and targeted influence at scale. In such a world, a person who cannot question, verify, slow down, and examine what they are being shown becomes more vulnerable.

This means discernment becomes a survival skill. The ability to notice when something is emotionally engineered, when evidence is weak, when confidence outruns truth, or when convenience is being used to bypass thought may prove extremely important.

The AI age will not only reward productive people. It will also reward those who can remain mentally independent inside environments designed to shape their attention and behavior.

Character Still Matters

Beneath all these specific skills lies something even deeper: character. Character includes honesty, steadiness, discipline, courage, humility, patience, responsibility, and the willingness to do what is right even when it is not convenient.

Some people assume that talk of character is vague or moralistic. In reality, character has practical consequences everywhere. A dishonest person with strong tools is more dangerous than a less capable honest person. A disciplined person can grow skill more reliably than a talented but unstable one. A courageous person can face change without collapsing into imitation or panic.

That is why the future will not be shaped only by what people know how to do. It will also be shaped by what kind of people they become while doing it.

Not Becoming a Weaker Version of the Machine

One of the great temptations of the AI age is to become a weaker version of the machine: faster on the surface, more scattered in attention, more dependent on convenience, more output-driven, and less inwardly serious. That temptation must be resisted.

The better path is not to mimic the machine, but to use it while becoming more clearly human. That means more capable of judgment, more grounded in responsibility, more disciplined in attention, more serious in communication, more trustworthy in relationships, and more thoughtful about consequences.

In other words, the future belongs not simply to those who have access to powerful tools, but to those who remain human enough to guide them well.

The age of AI does not reduce the skill question. It raises it. But it raises it in a different direction than many people assume. Technical fluency matters, but the most durable advantages may lie in judgment, communication, trust, adaptability, ethical reasoning, and character.

That leads naturally to the next question. If traditional work no longer remains the sole organizer of meaning, and if human life must be built on more than employability alone, then what should a serious life actually look like? That is where the next chapter begins.

Chapter 12

Education in the Age of Artificial Intelligence

Horizon Insight: When polished answers become easy, education must move from output to formation: attention, honest effort, clear thinking, communication, and judgment.

This chapter turns the skills question into a school question. If judgment, communication, curiosity, and honest process matter more, then classrooms and families must change how they define learning.

My teaching experience makes this subject especially important to me. I have seen students who could complete assignments but still struggle to explain why an idea mattered. I have also watched young children, including my own grandchildren, approach the world with natural curiosity before schooling and pressure begin to narrow their questions. AI can either deepen that curiosity or make it easier to avoid the struggle of learning.

Why Education Moves to the Center

A Grandparent's Stake in Education

I write about education not only as a former professor, but also as a grandfather. When I watch young children ask questions, I do not see a shortage of intelligence. I see an abundance of curiosity. The danger is that schooling, pressure, screens, and shortcut tools may gradually train that curiosity downward until learning becomes mainly performance.

AI makes this danger sharper, but it also creates an opportunity. A child who uses AI only to finish homework faster may become weaker. A child who uses AI to ask better questions, compare explanations, build projects, and explain ideas in their own words may become stronger. The difference will depend heavily on the adults around that child.

Education is not only the concern of schools. It is a family question. A grandparent watching a child use AI for homework is not merely watching a new tool enter the house. The grandparent is watching a question about character, curiosity, honesty, and future independence. Will the child use the tool to ask better questions, or to avoid thinking? Will the tool widen curiosity, or shrink patience? Will the child become stronger, or merely faster?

That is why education belongs near the practical center of this book. The AI age will not be handled well by policy alone, or by corporate innovation alone. It will be handled in kitchens, classrooms, libraries, colleges, workplaces, and family conversations where young people learn what kind of thinking is honorable and what kind of shortcut weakens the self.

Education is where human formation becomes personal. Every parent, teacher, and grandparent understands that a child is not only preparing for a test. A child is becoming a person. In the AI age, that distinction becomes even more important.

If artificial intelligence changes work, attention, judgment, and the structure of opportunity, then education can no longer remain a side issue. It moves to the center. Education is where a society prepares human beings not only to function in the world they inherit, but also to remain free, thoughtful, and capable of shaping that world well.

That is why the AI age forces us to rethink education more seriously than many people first realize. The issue is not simply whether schools should adopt new tools. The issue is what education must now become if powerful systems can increasingly provide answers, generate polished outputs, and weaken older assumptions about what counts as learning.

This chapter begins that discussion by asking what changes when intelligence itself becomes widely available as a tool.

The Limits of the Old Model

Much of modern schooling was shaped by the needs of an earlier age. It assumed information scarcity, standardized delivery, stable institutions, and a world in which success often depended on mastering known content and reproducing it reliably. That model had real strengths. It widened literacy, supported mass education, and helped societies build shared expectations.

But it also had limits. It often rewarded memorization more than judgment, compliance more than curiosity, and polished answers more than the visible process of thinking. Those weaknesses were present before AI. Artificial intelligence simply exposes them more sharply.

If a machine can generate essays, explanations, summaries, outlines, solutions, and even plausible class discussion language, then a system built too heavily around output alone becomes easier to game. The problem is not only cheating. The deeper problem is that the old model may no longer reveal clearly whether a student actually understands.

When Students Learn to Hate School

In my experience, most students — even many college students — do not actually love school. They may want good grades, credentials, or future opportunities, but that is not the same as loving learning. Too often, school is organized around memorizing material, passing exams, and performing for evaluation rather than exploring ideas and developing curiosity.

You can see the result in ordinary student life. Many students cannot wait for class to end, cannot wait for the semester to be over, and cannot wait for vacation. Something is fundamentally wrong when young people spend the bulk of their waking lives in school, yet experience that time mainly as something to get through.

I notice the contrast even more when I watch my own young grandchildren. Young children are naturally curious. They ask why, how, what if, and what happens next, not because there will be a test, but because the world itself is interesting to them. Yet somewhere along the path through school, too many young people lose that eagerness. By high school or college, the question often becomes, "Will this be on the exam?" That change should trouble us. It suggests that our current way of educating children can unintentionally train curiosity out of them at the very moment when lifelong curiosity is becoming one of the most important human strengths in the AI age.

Curiosity Is Like the Natural Voice

As a singer, I think about this loss through another analogy. A baby or very young child does not need to be taught how to project sound. When a baby cries, the whole body is involved. The breath is deep. The diaphragm is active. The sound carries. No one has explained vocal technique, posture, resonance, or breath support. The child simply uses the body naturally.

Then, as children grow older, they are taught to become quieter. Do not scream. Do not speak too loudly. Do not disturb the room. These lessons are socially necessary, of course. No family or classroom can function if every child cries out at full force all day. But something else often happens along the way. Many people gradually lose the natural connection between breath, body, and voice. By adulthood, when they try to sing -- especially classical singing, where the natural unamplified voice must carry through space -- they often discover that they no longer know how to use the body they were born with. Only a small percentage, through patient training and disciplined practice, recover that natural power in a mature and beautiful form.

Curiosity can suffer a similar fate. Young children ask questions with their whole being. Why? How? What if? What happens next? The world itself calls out to them. But if schooling trains them mainly to memorize, comply, prepare for tests, and ask only what will be on the exam, then the natural voice of curiosity grows softer. By the time many students reach high school or college, curiosity has not disappeared completely, but it has often been weakened. It no longer projects with the same force.

This is a serious educational loss because curiosity, once weakened, is hard to recover. Like the natural singing voice, it can sometimes be rebuilt through encouragement, practice, mentorship, and the right environment. But recovery is much harder than protection. A wise education should therefore not treat curiosity as a childhood decoration. It should treat curiosity as one of the central powers of human life.

In the age of AI, this matters even more. If students lose curiosity, AI may become only a shortcut machine. If students keep curiosity alive, AI can become a doorway into deeper questions, wider exploration, and more serious learning. The difference is not only technological. It is human.

This is one of the clearest signs that major educational renewal is needed. Properly used, AI may provide a rare opportunity to shift education away from rote performance and toward deeper inquiry, exploration, and genuine intellectual engagement. If we miss that opportunity, we may simply use a new technology to reinforce an already unhealthy model.

From Information Transfer to Human Formation

That is why education in the AI age should move beyond information transfer. Information still matters. Knowledge still matters. But the role of education must now be framed more broadly as human formation.

Formation includes judgment, curiosity, attention, discipline, honesty, resilience, communication, and the ability to think in the presence of abundant information. It means helping students not simply produce answers but become people capable of evaluating answers.

That shift is profound. Education can no longer justify itself mainly by saying that schools deliver information students cannot get elsewhere. Information is increasingly everywhere. The real value of education lies in helping students become the kind of people who can use information wisely.

How AI Tutoring Changes Learning

One of the most powerful new realities is that students now have access to something like a personalized tutor in their pocket. An AI system can explain a concept in multiple ways, answer follow-up questions, translate an explanation into simpler language, generate examples, create practice problems, and continue patiently for as long as the learner wishes.

This creates enormous opportunity. A student who is shy in class can still ask questions privately. A learner who needs repeated explanation can receive it without embarrassment. Someone working independently can explore ideas far beyond what a textbook alone provides.

But the same tool can also weaken learning if it is used mainly to avoid struggle. The difference lies in the philosophy of use. AI can deepen thought, or it can replace it. It can become a mentor-like guide, or a shortcut around formation. That distinction may become one of the most important educational questions of the next generation.

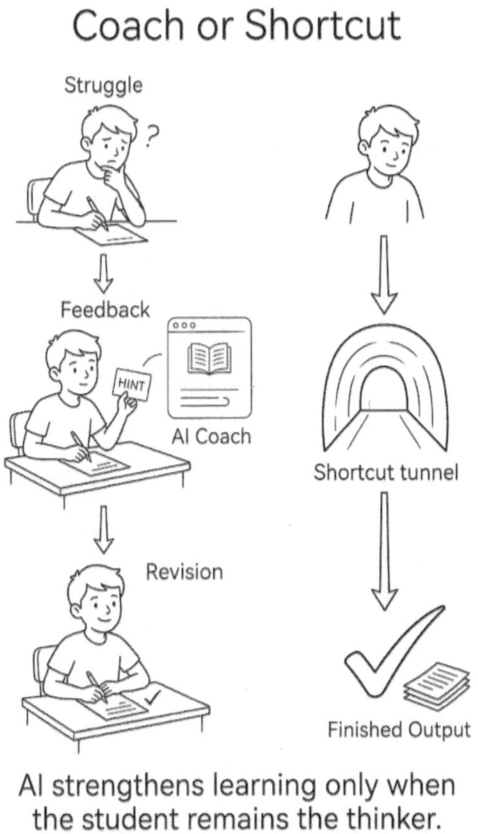

The reader should see the difference between AI as a coach that strengthens the learner and AI as a shortcut that weakens the mind.

The Teacher Becomes More Human

Some fear that AI in education means the teacher becomes less important. In reality, the opposite may often be true. If machines can assist with certain forms of explanation, drill, example generation, translation, and administrative burden, then the human role of the teacher becomes more visible.

A true teacher does more than deliver content. A teacher notices confusion, challenges laziness, awakens curiosity, models seriousness, strengthens confidence, and helps shape character. A teacher draws students into a way of seeing and living, not merely into a body of information.

That human role cannot simply be automated away. In fact, the more students are surrounded by fluent machine output, the more they may need adults who can model judgment, truthfulness, patience, attention, and intellectual integrity.

Assessment Must Change

If AI can generate respectable outputs quickly, then assessment must change as well. Schools can no longer rely too heavily on assignments that reveal only the final product while hiding the process. More emphasis may need to fall on oral explanation, iterative drafts, in-class reasoning, discussion, project work, version history, and visible traces of thought.

This is not merely a policing problem. It is an opportunity to improve education. A better assessment model can make learning more authentic by focusing less on polished surface performance and more on genuine understanding. Students should not only present answers. They should show how they got there, what they considered, what changed, and what they now understand more deeply.

In this sense, AI pressures schools toward a healthier educational honesty. It becomes harder to pretend that output alone equals learning.

The Ownership of Thought

One of the great educational questions of the AI age is whether students still experience ownership of thought. A person who turns to the machine too quickly may produce competent work without ever wrestling with the ideas long enough for them to become part of the self.

That matters because education is not just about public performance. It is about interior growth. A student must at some point say, “I see why this is true,” “I understand why this matters,” or “I can now think this through on my own.” If that moment is skipped, then the student may appear capable while remaining inwardly dependent.

So, the issue is not whether AI appears in education. It already has. The issue is whether students remain owners of thought or become mere editors of generated content.

Equity and the New Educational Divide

Artificial intelligence may also widen educational inequality if it is not handled well. Students with strong support, better devices, stable home environments, and adults who know how to guide AI use may benefit far more than students who lack those conditions. This creates the risk of a new divide: not only between students who have information and students who do not, but between students who know how to use AI to deepen learning and students who use it mainly to bypass learning.

That means education systems cannot simply tell students to “use AI responsibly” and assume the problem is solved. Schools will need to teach the philosophy of use explicitly. They will need to show how AI can support deeper questioning, stronger reasoning, and clearer ownership rather than becoming just another engine of shortcut culture.

Seen this way, the challenge of AI in education is also a challenge of guidance and equity.

Why This Is About More Than Schooling

Education in the age of AI is not only about schools. It is about the kind of human beings a society is trying to form. A civilization that teaches young people mainly to produce acceptable outputs will become fragile when machines can produce those outputs more easily. A civilization that teaches people to think, judge, learn continuously, communicate well, and remain morally serious will be stronger in the presence of powerful systems.

That is why this chapter is near the center of the book. Education is where a society decides whether the next generation will merely use intelligent tools or become capable of guiding them wisely.

College Still Matters - If It Changes

This same argument applies to higher education. Many people now question whether college is still worth the cost. They have a point if colleges remain unchanged. If universities continue to rely mainly on old lectures, old assignments, old exams, and the old assumption that a degree by itself proves readiness, then many graduates may leave with expensive credentials but without the capabilities that the AI age now requires.

The job market makes this question more serious. Some entry-level tasks that once helped young graduates begin a profession, such as first drafts, summaries, routine analysis, simple coding, administrative coordination, and standard reports, can now be assisted or partly automated by AI. If those beginner rungs become thinner, then young people need a different kind of preparation. They need to graduate not only with a transcript, but with judgment, communication skill, AI fluency, ethical awareness, project experience, and proof that they can learn, build, explain, and revise. Without that, they may be caught in the Entry-Level Catch-22: employers want experience, but students cannot get experience unless someone gives them a serious path into real work.

For that reason, colleges must bring more of the real world into education. Students need more case-style teaching, realistic scenarios, team projects, oral defense of their reasoning, employer-sponsored problems, internships, co-op programs, and feedback from people who actually work in the fields they hope to enter. Harvard Business School's case method is one well-known example of learning through concrete situations and practical judgment; that spirit should not remain limited to business schools. Engineering, computer science, healthcare, education, public policy, law, design, and the humanities can all help students connect knowledge with real-world judgment.

But the answer is not to give up on college. The answer is to demand a better college education. A university that teaches students how to think with AI, question AI, verify AI, and use AI responsibly can become more valuable than before. It can move beyond information delivery toward human formation: serious projects, visible process, oral defense, portfolios, mentorship, interdisciplinary breadth, and the human skills that machines cannot provide.

So, the better question is not simply, "Is college worth it?" The better question is, "What kind of college education is worth it now?" In the AI age, the value of college will depend less on the credential alone and more on the kind of thinker, worker, citizen, and human being the student becomes. Volume 4 of this book series develops this argument more fully because parents, teachers, mentors, and grandparents all need to help young people choose and use higher education wisely.

If education must now move from information transfer toward formation, then the next question is how learning itself must change across a whole lifetime. The AI age does not affect only students. It affects professionals, parents, older adults, and anyone who hopes to remain mentally alive in a changing world.

Chapter 13

Lifelong Learning in the Age of AI

> ***Horizon Insight: Learning is no longer a phase before adult life. It becomes a lifelong expression of dignity, adaptability, curiosity, and renewal.***

This chapter moves beyond school. AI does not only challenge students; it challenges workers, parents, grandparents, retirees, and citizens to remain teachable across the whole span of life.

At my age, I no longer think of lifelong learning as a slogan for young people. It is a discipline for every generation. When I learn new AI tools, prepare lectures, or experiment with demonstrations, I am also reminding myself that staying curious is one way older adults remain part of the future rather than standing outside it.

Why Education Can No Longer End Early

Older Adults as Learners, Not Bystanders

I have found that older adults often learn best when AI is connected to a real need rather than presented as a technical spectacle. Explain a confusing letter. Prepare questions for a doctor. Compare two travel options. Summarize a family document. Translate a message. These small uses matter because they turn AI from a frightening abstraction into a practical helper.

The dignity of learning later in life is that it proves the future does not belong only to the young. We may not learn every tool as quickly, and we do not need to chase every novelty. But we can remain present, curious, and useful. That presence itself is a gift to the family.

One of the most encouraging things I see in community AI lectures is the moment when an older adult moves from embarrassment to curiosity. At first, the person may say, "I am not technical." Then, after seeing a confusing letter explained or a medical question organized, the expression changes. The tool is no longer an alien object. It becomes a possible helper. That small shift matters because it restores participation.

The point is not that older adults must chase every new platform. They should not. The point is that they should remain learners. Learning keeps a person connected to the present and to the younger generation. It protects independence. It also gives older adults a stronger voice in family conversations about AI, because wisdom without current understanding can be ignored too easily, while current understanding without wisdom can be reckless.

Many older adults know this truth from experience: life keeps teaching long after school ends. The AI age simply makes that lifelong lesson more visible and more urgent.

For much of modern life, education has often been treated as a front-loaded stage. A person studies when young, earns credentials, enters a profession, and then spends decades applying what they learned with only periodic updating along the way. That model was never perfect, but it fit a world that changed more slowly.

The age of artificial intelligence places that model under increasing pressure. When tools, workflows, expectations, and even entire categories of work shift more quickly, education can no longer be treated as something completed early and simply carried forward unchanged. Learning must become more continuous.

That is why lifelong learning now matters at a much deeper level. It is not just a strategy for getting ahead. It becomes part of what it means to remain capable, dignified, and mentally alive in a changing world.

Retirement should not mean retirement from learning. One mistake I often hear after my AI lectures is the thought, "I am retired, so this no longer concerns me." But the AI age will enter retirement life too. It will affect medical information, financial decisions, travel, communication with family, fraud prevention, home support, and the way older adults remain connected to the young.

The deeper issue is not employment. It is participation. A retired person who refuses to learn anything about AI may not lose a job, but may gradually lose confidence, independence, and influence. That is a different kind of obsolescence. The goal is not to chase every new tool. The goal is to remain teachable enough to stay present in the world as it changes.

More Than Reskilling

When people hear the phrase lifelong learning, they often think first of reskilling: learning a new tool, adapting to a new system, updating a professional method, or adding a marketable competency. All of that matters. But lifelong learning in the AI age must mean more than periodic vocational repair.

A person who learns only enough to avoid immediate obsolescence may remain employable for a time, but still become inwardly passive. The deeper aim of lifelong learning is not merely to survive economically. It is to remain intellectually awake, teachable, and open to growth.

This matters because the challenge of the AI era is not only technical change. It is also the temptation to hand more and more of one's thinking over to convenient systems. Lifelong learning resists that temptation by keeping the mind in active motion.

The Half-Life of Knowledge

One reason lifelong learning becomes more necessary is that the half-life of practical knowledge grows shorter in many fields. Methods that once remained stable for long periods may now change more quickly. Software changes. Platforms change. Tools change. Expectations change. Workflows change. Even the basic assumptions of a profession may begin to shift when intelligent systems alter what humans are expected to do themselves.

This does not mean all old knowledge becomes useless. On the contrary, foundational understanding often becomes even more valuable because it helps people interpret change rather than merely react to it. But it does mean that no one can assume a single burst of education in youth will be sufficient for an entire adult life.

The person who stops learning may not notice the decline at first. But over time, distance grows between that person and the frontier of how the world is actually moving. Eventually, what felt like stability may reveal itself as stagnation.

Teachability as a Human Strength

One of the most underestimated human strengths is teachability. Teachability means more than intelligence. It means the willingness to learn, the humility to admit what one does not yet understand, the courage to experiment without immediate mastery, and the discipline to keep growing after the ego would prefer to stop.

This matters in the AI age because many people will be tempted to retreat into one of two errors. Some will become defensive and refuse to engage new tools at all. Others will pretend fluency without doing the harder work of real understanding. Teachability avoids both mistakes. It allows a person to remain honest while still remaining open.

A teachable person does not become less serious with age. They become more serious about staying alive to reality.

AI as a Partner in Learning

Artificial intelligence can also become a powerful aid to lifelong learning. An adult learner can ask for explanations, request examples, translate technical language into clearer language, test understanding through dialogue, and explore unfamiliar fields without waiting for a formal class. This lowers the barrier to continued learning in ways that can be genuinely liberating.

A middle-aged professional can learn the basics of a new field faster than before. An older adult can ask patient questions without embarrassment. A parent can explore a topic deeply while balancing family demands. Someone who feels rusty can re-enter learning with less intimidation.

Used well, AI can function as a patient tutor, a translator of complexity, and a guide into new domains of knowledge. It can make lifelong learning more accessible than earlier generations could easily imagine.

Learning can continue across the whole life span.

The Danger of Shallow Pseudo-Learning

Yet AI also creates a new danger: shallow pseudo-learning. A person may feel informed because they can quickly generate explanations, summaries, and answers, while never really building durable understanding. They may confuse access with mastery and familiarity with knowledge.

For that reason, the philosophy of use matters so much. If AI is used to deepen thought, ask better questions, test understanding, and move gradually toward genuine ownership, it can be a remarkable aid. If it is used mainly to avoid mental effort, then it may produce the illusion of learning without its substance.

The danger is subtle because the person may not immediately realize what is missing. They can speak more fluently, produce more output, and seem more informed. But when the moment comes that requires independent judgment, the borrowed surface may collapse.

Learning as Dignity, Not Only Utility

There is another reason lifelong learning matters: it protects dignity. A person who continues learning is less likely to feel simply left behind by the age they are living through. They remain in conversation with the present rather than being pushed to the margin of it.

This is especially important for older adults and for people who fear that rapid change belongs only to the young. The ability to keep learning is not a concession to decline. It is one of the clearest signs of continuing vitality. A mind that remains curious and teachable remains more alive.

That is why lifelong learning should not be framed only as market adaptation. It is also a defense against internal shrinking. It helps people remain participants in reality rather than spectators of a world they no longer understand.

Learning and Identity

A person who keeps learning also develops a healthier identity. Instead of defining the self too narrowly by a fixed role, they begin to understand life as ongoing formation.

They remain capable of change without losing inner continuity. They can say, in effect, “I am still becoming.”

This is particularly valuable in the AI age because many old identities tied to stable professional scripts may weaken. Learning helps create a more resilient self — a self not trapped by one role or one past success.

Seen this way, lifelong learning is not only about new information. It is about a new way of inhabiting adulthood: less fixed, less brittle, more open, and more durable.

A Culture of Continuous Renewal

If lifelong learning is truly necessary, then societies will need more than good advice. They will need cultures, institutions, and norms that support continuous renewal. Employers may need to make room for real retraining rather than simply demanding instant adaptation. Communities may need to create more accessible learning pathways. Schools and universities may need to think of themselves not only as early-life institutions, but as partners across the life course.

This matters because the burden of adaptation should not fall on isolated individuals alone. A humane society does not tell people to update endlessly while providing no structure for how to do it. If AI is a civilizational transition, then continuous learning must become a civilizational norm as well.

Why This Chapter Matters

Lifelong learning is one of the clearest ways human beings remain active rather than passive in the AI age. It resists the temptation to outsource too much thought. It protects dignity. It supports adaptability. It builds resilience. And it reminds people that growth does not belong only to childhood or formal schooling.

In a world of increasingly powerful tools, the person who keeps learning remains harder to marginalize and harder to manipulate. They remain capable of meeting the future as a participant rather than a bystander.

If education can no longer be understood merely as a youthful stage, and if lifelong learning becomes part of adult dignity and freedom, then the deepest educational question still remains: what is education really for?

That is the question of the next chapter.

Chapter 14

What Is Education Really For?

Horizon Insight: Education is not only job preparation. Its deeper purpose is to form free, thoughtful, responsible human beings who can live wisely with powerful tools.

This is not another skills chapter. It asks the deeper purpose question: what kind of person should education form when machines can supply answers but cannot supply wisdom?

If education is only training for the next task, AI will keep making parts of it look obsolete. But if education is the formation of judgment, character, curiosity, communication, and responsibility, then it becomes more important, not less. This is the message I would give to students, parents, teachers, and grandparents alike.

The Deeper Educational Question

The Student Who Can Explain

In the AI age, I would increasingly trust the student who can explain the work more than the student who merely submits polished work. Explanation reveals ownership. It shows whether the idea has passed through the student's own mind or only through the machine. A student who can defend a choice, revise an argument, answer follow-up questions, and admit uncertainty is building real strength.

This is also why oral discussion, drafts, projects, and version history matter. They return education to process. They make visible the struggle by which understanding becomes personal. Without that struggle, AI may produce good-looking surfaces while leaving the mind underneath less formed.

In the AI age, one of the simplest tests of real education may be this: can the student explain the work in their own words? Not recite a generated paragraph. Not point to a polished submission. Explain it. Defend it. Connect it. Admit what is uncertain. Show how the idea changed through revision. When a student can do that, the work has entered the mind rather than merely passing through a machine.

This test is old, but AI makes it urgent. For centuries, good teachers have known that the final answer is not enough. Now the whole culture must remember it. Education that cannot distinguish between produced output and owned understanding will become easier to fool. Education that asks students to show process, reasoning, voice, and judgment will become more valuable than before.

The previous two chapters argued that education in the age of artificial intelligence should move beyond old assumptions. It can no longer be understood mainly as information transfer, and it can no longer be confined to a youthful stage followed by static adulthood. Those shifts matter. But they still lead to a deeper question: what is education really for?

This question has always mattered, but the AI age makes it harder to postpone. If machines can increasingly generate answers, explain concepts, and assist with many forms of output, then education cannot justify itself merely by saying that it delivers information or prepares people for narrow functions inside the labor market. It must be able to say what kind of human being it is trying to form.

That is why this chapter stands near the philosophical center of the book. Education is never only about skills. It is always, whether openly admitted or not, about the formation of persons.

Education Is Not Only Job Preparation

Job preparation remains necessary, but it is too small to carry the whole meaning of education. In the AI age, schools must prepare people not only to earn, but to judge, participate, explain, care, and remain free.

Modern societies often speak about education in utilitarian terms. We ask whether it prepares students for jobs, whether it matches market demand, whether it raises productivity, whether it improves national competitiveness. These questions are not trivial. Human beings do need to make a living, and societies do need competent workers.

But if education is reduced entirely to job preparation, it becomes too small for the human task. A person is more than a future employee. A human being must also become a thinker, a citizen, a moral agent, a steward of attention, a participant in community, and eventually a guide to others.

This is not an abstract point for me. In my years of teaching college and graduate students, I often saw students who were bright and diligent, but too narrowly trained. Some could handle technical material, yet struggled to write well, present ideas persuasively, read deeply, or connect their subject to history, culture, ethics, business, or world affairs. In the AI age, that narrowness becomes more dangerous. If machines can help produce technical output, then the human advantage increasingly lies in breadth, judgment, communication, and the ability to see a problem in its larger human setting.

The AI age intensifies this truth because technical preparation alone no longer guarantees maturity. A society may produce highly trained people who know how to use powerful tools and yet remain shallow in judgment, weak in character, and vulnerable to manipulation. That is not a fully educated society. It is merely a technically armed one.

Education as Formation

A richer view of education begins with the idea of formation. Formation means the shaping of mind, character, habits, and perception. It asks not only what a student knows, but what kind of person the student is becoming through the process of learning.

A formed person can examine a claim rather than merely repeat it. They can distinguish confidence from truth. They can tolerate complexity without collapsing into slogans. They can attend, reflect, revise, and think through consequences. They can resist being carried entirely by the emotional currents of the moment.

This kind of formation matters even more when machines can generate polished responses quickly. If students learn only to produce output, then AI may outshine them at the very thing they were trained to do. But if students are formed into people who can interpret, judge, question, and connect knowledge to reality, then education retains its deepest value.

A school system that leaves students counting the minutes until class ends and the days until vacation has failed at one of its deepest purposes. Education should awaken curiosity and intellectual life, not merely train students to endure memorization and performance.

As the earlier education chapter argued through the analogy of the natural singing voice, curiosity should not be treated as something we can easily recover after years of neglect. Education should protect it early, strengthen it deliberately, and mature it into disciplined inquiry.

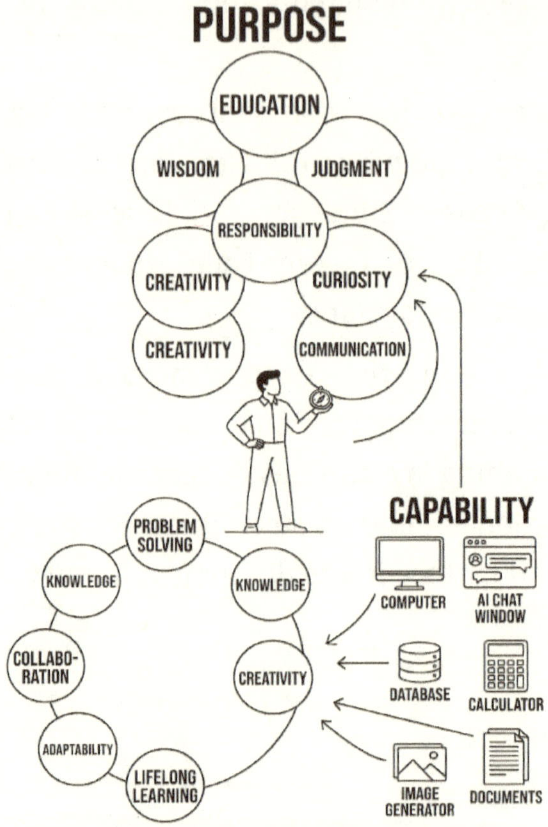

Education is not preparation for work.
It is formation of thoughtful beings.

Education forms people, not only workers.

Education and Freedom

Education is also preparation for freedom. A free human being is not merely someone who has options. A free person is someone capable of choosing well among options. That requires more than technical knowledge. It requires judgment, discipline, memory, moral seriousness, and the ability to resist manipulation.

That is why education matters politically as well as personally. A society of people who cannot evaluate claims, examine evidence, endure disagreement, or think beyond stimulation becomes vulnerable. It may remain technologically advanced while becoming civically fragile.

Artificial intelligence raises this issue sharply because it increases the supply of persuasive language, targeted messaging, synthetic media, and plausible but misleading explanation. In such a world, the uneducated mind is not merely uninformed. It is exposed. Education becomes one of the main defenses of freedom.

Critical Thinking Is Not a Slogan

Many educators speak of critical thinking, but often in a vague way. In the age of AI, critical thinking must become more concrete. It means asking what assumptions are hidden in an argument, what evidence is missing, what alternatives were ignored, what incentives may shape a claim, and what consequences may follow if the claim is accepted.

Critical thinking also includes the ability to slow down. In a world of rapid outputs, one of the most important intellectual acts may be the refusal to accept the first fluent answer as final. A student must learn to interrogate, compare, reframe, and test.

Again and again, in teaching, I have seen that the most valuable moments in education are not when a student repeats the correct answer, but when something deeper becomes visible — when the student really sees why something is true, why it matters, or why a previous assumption was incomplete.

I was struck by this recently in a conversation with a highly accomplished New York attorney. She told me that some of her junior law clerks were already using AI, but not wisely. They could produce faster drafts, but they often did not internalize the essence of the material they generated. As a result, they were not very useful: they could not clearly explain their findings to her, and worse, they could not explain them well to clients. She told me, quite bluntly, that those would be the clerks laid off first.

That conversation stayed with me because it revealed something important. The problem is not whether young professionals use AI. The problem is whether they use it to deepen their thinking or to bypass it. When AI-generated material never becomes part of a person's own understanding, the surface may improve while the human being underneath becomes weaker.

This is one reason AI may become a strangely clarifying force in education. It exposes how often schooling rewards answer production without requiring genuine examination. Now that answers can be produced cheaply, the ability to think critically about answers becomes more central.

The Humanities Become More Important, Not Less

Some people assume that in a highly technological age, the humanities become less relevant. I believe the opposite may be true. History, philosophy, literature, religion, art, and the study of human suffering and aspiration may become more important precisely because the machine can do so much in more technical domains.

The humanities do not merely provide information. They enlarge moral imagination. They help people confront tragedy, ambiguity, power, beauty, betrayal, courage, mortality, and the long human struggle to live meaningfully. They place the present moment within a wider human story.

This matters because the AI age will tempt people to think that capability is enough. The humanities remind us that capability without wisdom can become destructive, and that intelligence without moral depth is not the same as civilization.

Education and Citizenship

A serious education also prepares people for citizenship. Citizenship does not only mean voting or knowing facts about government. It means being able to participate in a shared public world with enough discipline and honesty to keep that world from collapsing into manipulation, tribalism, and spectacle.

The AI age may make this more difficult. If synthetic media, persuasive systems, targeted messaging, and emotionally optimized content become more widespread, then citizens will need stronger habits of verification and a deeper commitment to truth. Without that, democracy itself may become easier to destabilize.

That is why education cannot be understood only as private advancement. It is also a public safeguard. A nation that neglects the education of judgment may gain new tools while losing the civic maturity needed to use them well.

Beyond Credentialism

Another distortion that AI may expose is credentialism: the tendency to treat education mainly as a process of collecting symbols that signal social value rather than becoming more deeply educated. Credentials can be useful. They often reflect real effort and real competence. But they can also become substitutes for formation when institutions care more about measurable signals than about actual understanding.

If AI makes it easier to produce respectable outputs, then superficial credentialism may become even more fragile. What will matter more over time is not merely whether a person passed through a system, but whether they can actually think, communicate, judge, create, and act responsibly in the real world.

This may be one of the hidden gifts of the AI age. It may force education to confront whether it is producing genuinely formed human beings or merely credentialed performers.

The Goal Is a Fully Educated Human Being

At its highest level, education aims at a whole person, not a polished performer. Such a person can learn continuously, examine claims, communicate honestly, recognize moral stakes, appreciate beauty, and use freedom responsibly.

This is why education remains central even when AI becomes powerful. The machine can help produce an answer. Education must help form the person who knows what kind of answer is worth seeking, what kind of evidence deserves trust, and what kind of life deserves devotion.

Why This Chapter Matters for the Whole Book

This chapter matters because education is where the whole argument of the book converges. The economy matters. Work matters. Meaning matters. Generational advice matters. But all of these depend in part on whether societies know how to form human beings equal to the tools they are creating.

If AI becomes one of the most powerful technologies in human history, then education becomes one of the most important human responses. Not because it can stop the technology, but because it can shape the kinds of people who will live with it.

With that foundation in place, the book can now return to the deeper meaning questions with more practical grounding. If education forms judgment, character, and freedom, then the next question is what human beings should do with those capacities when work itself is transformed.

The next section therefore moves from education as formation to work as meaning. Once education has prepared human beings to think, judge, and act freely, the book must ask what those capacities are for when AI begins to transform the old structure of work itself.

Part III Summary

Education as Formation in the AI Age

Part III has argued that education in the age of AI cannot be reduced to information transfer, job training, or credential collection. If machines can increasingly generate answers, then education must form the person who can ask better questions, judge the quality of answers, and carry responsibility for how knowledge is used.

That means schools, families, and adult learners must protect curiosity, cultivate explanation, strengthen communication, and build habits of verification. The educated person of the AI age is not simply someone who can produce polished output, but someone who can think with ownership, speak with clarity, and remain free in the presence of powerful tools.

Part IV Introduction

Work, Meaning, and the Human Need to Contribute

With education understood as formation, the book is ready to turn to the next human question: work. If AI changes what people do, it will also change how people experience usefulness, contribution, dignity, and meaning.

Work has never been only a way to earn money. It has also given people structure, responsibility, identity, competence, recognition, and the feeling of being needed. That is why the AI age raises a deeper question than whether certain tasks will be automated. It asks what happens to human beings when the old relationship between labor, income, dignity, and purpose begins to change.

This part therefore explores work as a human institution, not merely an economic function. It asks why people have always worked, why meaningless life is dangerous, what may replace work as the central organizing force of identity, and how a more abundant society might become a meaning society rather than a society of passive consumption.

The goal is not to romanticize work or deny the hardship many people experience inside it. The goal is to understand what work has provided psychologically and socially, so that if AI changes the structure of work, society does not accidentally remove the human supports that work once supplied.

Chapter 15

Why Humans Have Always Worked: The Meaning of Work

Horizon Insight: Work has never been only about wages. It gives income, structure, competence, recognition, and the human feeling of being needed.

This question is personal to me because I have seen work from many sides: the discipline of research, the responsibility of teaching, the pressure of running a company, and the human need for contribution after retirement. Work is not only income. For many people, it is rhythm, recognition, usefulness, and a way to feel connected to others.

The Deeper Question Beneath Work

Jobs are visible on the outside, but work reaches into the inside of a person. It gives rhythm to the day, responsibility to effort, recognition from others, and the quiet feeling that one is needed.

The question beneath the AI economy is therefore not only what people will do for wages. It is what happens when a civilization weakens one of the main structures through which people have found income, discipline, usefulness, and identity.

That question is deeper than it first appears. If work were only a way of earning money, then the answer to technological abundance would be simple. If machines can produce more, then human beings could simply consume more and relax. But human life has never been organized only around income. Work has always served functions that go beyond wages.

That is why the meaning crisis of the AI age cannot be understood merely as an economic problem. It is also a human problem. If the structure of work weakens, then something more than income may be shaken.

Work Long Before Salaries

Work as the Place Where the Self Meets Reality

Work has a way of revealing the self because reality pushes back. A plan fails. A customer is unhappy. A student does not understand. A program has a bug. A team member misunderstands the assignment. In those moments, a person learns patience, humility, persistence, and responsibility. These are not decorative virtues. They are the traits by which adults become reliable.

This is why I worry about a future that removes too much friction without replacing it with higher forms of responsibility. The human being is not formed by convenience alone. We are formed by meeting reality, carrying obligations, and discovering that our effort can matter to someone else.

One reason work matters is that it puts the self in contact with reality. A plan sounds good until the customer responds. A lesson sounds clear until the student remains confused. A business strategy sounds elegant until the market resists. A musical phrase looks simple on the page until breath, nerves, memory, and audience presence test it. Work reveals the distance between intention and reality.

That is why work has formed human beings for so long. It disciplines fantasy. It teaches consequence. It forces adjustment. If AI removes some burdens, that can be good, but human beings will still need places where intention meets resistance and where effort becomes character. A life with no such places may become smooth on the surface and weak underneath.

Human beings worked long before there were salaries, résumés, corporations, or formal labor markets. Early humans gathered food, built shelter, cared for children, made tools, protected communities, and maintained the practical conditions of life. None of this resembled the modern office, yet it was still work.

This matters because it reminds us that work is older than the modern economy. It is rooted in the human condition itself. To work is to engage reality, to respond to need, to shape the world, and to take part in sustaining life. Even in the absence of money, human beings still labored because life itself required effort, cooperation, and contribution.

The modern world changed the form of work, but not its deeper significance. We still experience work not only as a transaction, but as a way of inhabiting life with seriousness.

The Three Pillars of Work

Pillar	What It Provides	AI-Age Pressure	Human Response
Income	Material stability and independence	Routine labor may be compressed	Protect dignity through new opportunities and support
Structure	Rhythm, obligation, and useful discipline	More free time may become shapeless	Build chosen routines, learning, and service
Meaning	Recognition, contribution, and being needed	Utility may be automated or devalued	Move upward toward care, wisdom, and legacy

One useful way to understand work is to see it as resting on three pillars:

1. **Income.** Work provides the material basis for survival, stability, independence, and family responsibility.

2. **Structure.** Work organizes time. It gives shape to the day, the week, and the year.

3. **Meaning.** Work gives people the feeling that they are needed and that their effort matters to someone beyond themselves.

The age of AI matters because it may put pressure on all three pillars at once.

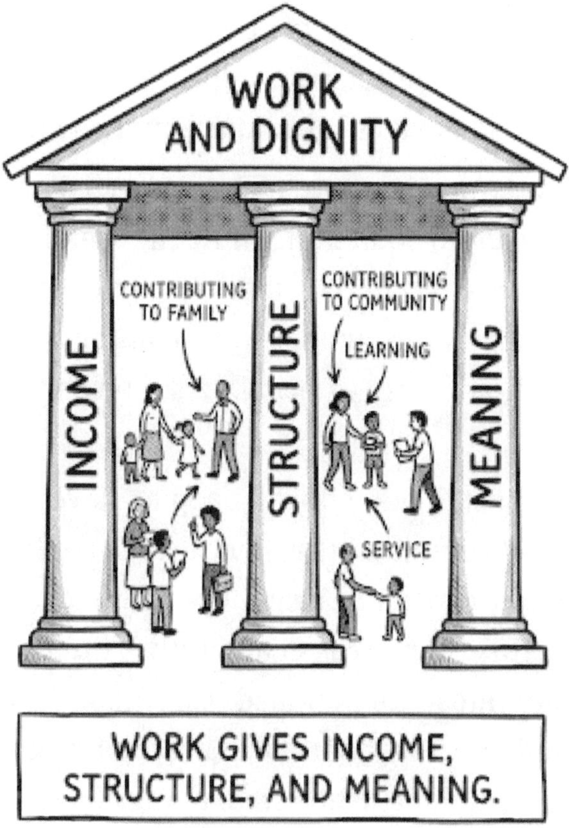

Utility and Significance

AI may increase utility by performing tasks faster, but significance comes from trust, care, wisdom, service, and human presence. As machines provide more utility, people must deepen the forms of contribution that make life weighty and personal.

Work as Participation in Reality

Human beings do not flourish by consuming alone. They also need to participate. To work is, in part, to participate in reality rather than merely drift through it. A person faces resistance, solves problems, improves skill, encounters limits, adapts, and leaves some mark on the world.

That is why work often feels connected to dignity. A person who can do something difficult, useful, or beautiful often feels more grounded than a person whose life is organized entirely around passive comfort. The value of work here is not merely external. It is internal. It shapes the worker.

Through sustained effort, people develop patience, discipline, self-command, resilience, and confidence. They learn what they can bear, what they can build, and how they can matter. This is one reason any future discussion of post-work society must be handled carefully. If we remove structure and effort without replacing their human function, we may create comfort while weakening character.

Why Being Needed Matters

One of the deepest human needs is the need to feel needed. A person who experiences themselves as unnecessary does not merely lose income. They may begin to lose orientation. They may ask whether their skills matter, whether their experience still counts, and whether they still have a recognized place in the world.

Work has long been one of the main ways society answers that question. A role, even an imperfect one, tells a person that someone depends on them. Their contribution may be modest, but it is visible. Their effort is tied to consequence.

That is why the displacement of work can cut so deeply. It is not simply the loss of a task. It can feel like the loss of social recognition. The AI age may intensify this risk if people begin to sense that systems can perform more of what they once believed made them valuable.

The Discipline of Effort

Human beings are not made better only by ease. Difficulty often plays a formative role. Work asks people to show up when they are tired, to continue when something is frustrating, to collaborate with others, and to do what is necessary before what is pleasant.

This discipline is not glamorous, but it is civilizationally important. A society is sustained not only by brilliance, but by reliability. The person who keeps going, who carries a burden, who meets an obligation, and who does the uncelebrated work that others depend on is helping to hold the world together.

If AI reduces the need for some of this effort, that may in some cases be a gift. There is no virtue in pointless drudgery. But the disappearance of effort is not automatically a gift if it removes one of the main ways human beings build seriousness. The real issue is not whether some labor becomes unnecessary. It is whether human beings still choose forms of effort that build them into stronger persons.

Work and Participation in Ordinary Life

Over the years, I have been reminded of this truth in many ordinary settings. A student who finally understands an idea, an employee who grows into responsibility, a customer who feels heard rather than processed, and a family member who receives steady care all reveal the same truth: human beings want to participate, not merely be maintained.

These moments are moving not only because music brings back memory, but because they reveal something about the human condition. People do not want only to be maintained. They want to join in. They want to respond. They want to feel that something still passes through them into the world.

That same impulse helps explain why work matters. Human beings want to take part in life, not merely observe it from a distance. They want their effort, voice, and presence to count for something.

Work and Identity

Modern societies have often tied identity very tightly to occupation. One of the first questions people ask is, "What do you do?" That question sounds practical, but it carries enormous psychological weight. It assumes that work is one of the clearest ways to define a life.

There is some truth in that. Work often becomes a site of mastery, purpose, and responsibility. But there is also danger in tying the self too narrowly to the role. If a role changes or disappears, the whole self may feel threatened.

This is one reason the AI transition may be emotionally difficult even when people remain materially secure. If identity has been built too narrowly around a professional function, then technological disruption may feel like a challenge not only to livelihood, but to selfhood. The human being begins to ask: if I no longer do this, who am I?

That question will grow more important as the book moves forward. But before we reach it fully, it is necessary to see why work held such power over identity in the first place. It held that power because work was never merely economic. It was one of the main ways people joined their inner life to the outer world.

Why This Matters in the Age of AI

Artificial intelligence forces this question into the open. If machines increasingly take over parts of production, analysis, drafting, scheduling, transportation, and perhaps even many forms of care support, then human beings may have more relief from necessity. That could be good. But relief from necessity is not the same as fulfillment.

If a society reduces the role of work without understanding why work mattered, it may produce a strange new kind of poverty — not a poverty of goods, but a poverty of significance, structure, and lived purpose. People may have more assistance and more convenience while feeling less needed.

That is why the next chapters in this section turn from work itself to meaning more directly. We must ask not only what work paid for, but what work gave. Only then can we begin to imagine what a meaningful human future would require if the old labor structure changes.

Human beings have always worked because work has always done more than generate income. It has joined people to necessity, responsibility, structure, contribution, and reality itself. It has helped form character and sustain dignity. It has answered, however imperfectly, the human desire to be needed.

If artificial intelligence weakens the old centrality of work, then the question is not simply what jobs disappear. The deeper question is what happens to meaning. That is where the next chapter begins.

Chapter 16

The Danger of a Meaningless Life

Horizon Insight: Comfort alone does not create purpose. If AI gives society more freedom from necessity, human beings must still choose effort, service, relationship, and responsibility.

When I think about the danger of a meaningless life, I do not think first about theory. I think about people whose days have lost rhythm, whose abilities have been underused, or whose experience is no longer invited into the life of others. Human dignity is strengthened when a person still has some way to give, respond, teach, care, create, or bless the lives around them.

When Comfort and Emptiness Grow Together

This chapter is one of the most important warnings in the book because it touches an old human truth. Ease is good when it relieves suffering, but ease alone does not tell a person why to get up in the morning.

A technological civilization can become more comfortable without becoming more meaningful. This is one of the deepest dangers of the AI age. If artificial intelligence reduces friction, automates more tasks, and expands convenience across daily life, many people may at first experience this as progress. Some of it will indeed be progress. But convenience alone does not tell us whether a life is going well.

Human beings do not suffer only from material deprivation. They also suffer from emptiness, drift, and the quiet erosion of purpose. A person may have more assistance, more entertainment, and more ease, yet still feel that something central is missing. If work weakens, if discipline weakens, and if life becomes organized increasingly around passive consumption, then abundance may coexist with inner poverty.

That is why this chapter matters. The question is not only whether AI can make life easier. The deeper question is whether human beings will still know what to do with life when effort is no longer imposed in the old way.

Meaning Does Not Arrive Automatically

The Danger of a Quietly Empty Abundance

The most dangerous kind of emptiness may not look dramatic. It may look comfortable. The person has entertainment, convenience, delivery, generated companionship, instant answers, and fewer obvious hardships, yet still feels that the day has no weight. This is not a failure of technology alone. It is a failure of orientation. Tools can make life easier, but they cannot decide what the ease is for.

A society moving toward AI abundance must therefore teach people how to use freedom. That may sound strange, but freedom is not automatic maturity. Freedom requires form: responsibilities, promises, relationships, service, craft, learning, worship or reflection for those who practice it, and some commitment to goods larger than appetite. Without form, freedom can dissolve into drift.

One of the most dangerous assumptions in modern culture is that if suffering is reduced and options are increased, meaning will somehow take care of itself. That is not how human beings work. Meaning is not the automatic byproduct of comfort. It usually arises through orientation, commitment, responsibility, love, service, sacrifice, growth, and chosen effort.

That is why abundance alone cannot solve the human problem. It can relieve hardship, and that matters greatly. But once basic hardship is reduced, another question returns with even greater force: what is this life for?

The AI age may intensify that question because it may weaken older structures that once answered it imperfectly but powerfully. Work gave routine. Scarcity imposed effort. Necessity organized the day. Those things often felt burdensome, yet they also gave life shape. If they weaken, then human beings must supply shape more consciously than before.

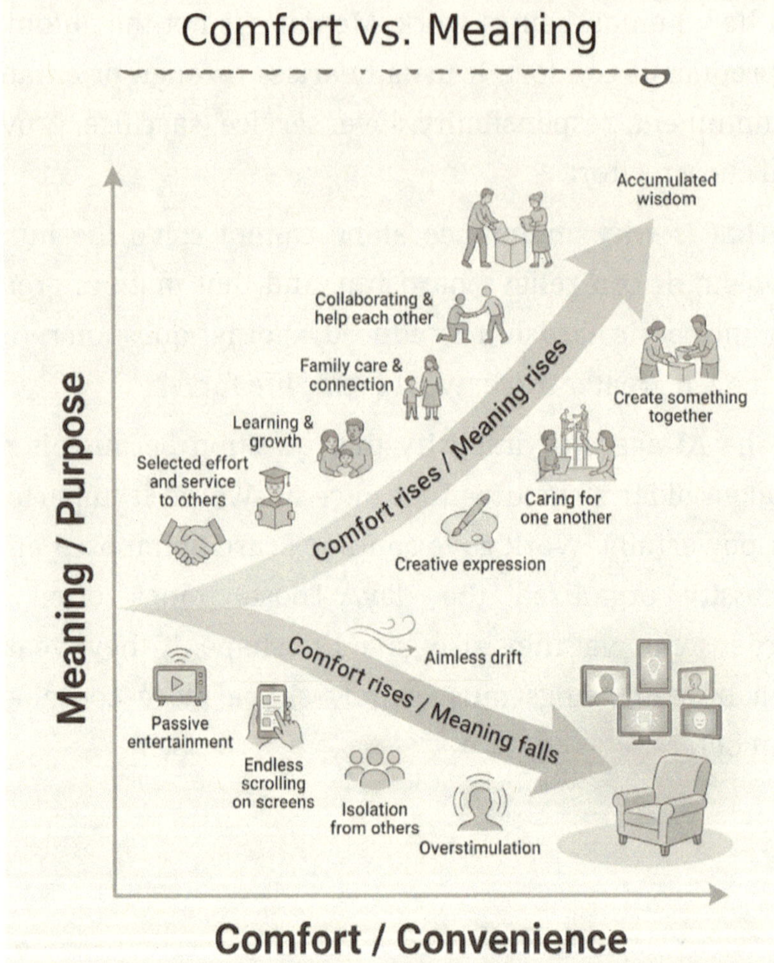

A richer life does not automatically follow from an easier life.

Comfort alone does not create meaning; human beings still need purpose, effort, and connection.

The Vacuum Left by Lost Structure

A life without structure does not remain neutral for long. If a person is no longer held by meaningful routines, obligations, responsibilities, and forms of effort, the empty space does not simply remain open and calm. It is usually filled by distraction, passivity, compulsive stimulation, or low-grade despair.

For that reason the weakening of work can be dangerous even when material conditions improve. A person may have enough to eat, enough entertainment, and enough digital convenience, yet still feel internally disorganized. The day loses weight. Time becomes harder to inhabit. The future becomes harder to shape.

Human beings need more than relief from pressure. They need a reason to rise, act, care, and continue. When that reason weakens, the whole inner structure of life can soften.

Pleasure Is Not Purpose

The temptation is to imagine that if AI produces more abundance, then people can simply enjoy themselves more. There is nothing wrong with pleasure in its place. Rest, beauty, delight, humor, and recreation are real goods. But pleasure is not purpose.

A life organized mainly around stimulation and comfort eventually becomes thin. Consumption can occupy attention for a while, but it rarely satisfies the deeper hunger of the human being. A person still wants to become someone, not merely experience things. They still want their life to count in relation to others, to truth, and to something larger than passing enjoyment.

This is one reason a civilization that confuses entertainment with fulfillment becomes fragile. It may look prosperous on the surface while growing weaker inwardly.

The Risk of Passive Consumption

Artificial intelligence may increase this danger because it can make passive consumption easier than ever. Content can be generated instantly. Companionship can be simulated. Digital systems can anticipate preferences, tailor stimulation, and reduce the need to struggle with boredom, ambiguity, or silence. A person may increasingly live inside a world of smooth response and endless personalization.

But human beings are not strengthened mainly by being endlessly catered to. They are strengthened by reality, by resistance, by responsibility, and by the discipline of engaging what does not instantly conform to desire.

If AI is used mainly to create a frictionless consumer existence, then it may deepen one of the great weaknesses of late modern life: the drift toward passivity. A society of highly entertained but inwardly unformed people is not a flourishing society, no matter how advanced its tools become.

The Erosion of Chosen Effort

Meaning often grows where effort is chosen rather than merely imposed. A person cares for a child, learns a craft, builds a relationship, creates something beautiful, serves a community, tends to a parent, keeps a promise, or carries a responsibility. These actions are meaningful not because they are easy, but because they require something of the self.

If AI removes too much difficulty without helping people redirect effort toward higher forms of contribution, then one of the conditions of meaning may be weakened. The machine will have relieved burden, but the person may not yet know what to do with the freedom that remains.

For that reason the issue is not whether difficulty should be preserved for its own sake. The issue is whether human beings will still choose forms of disciplined effort that shape character, connect them to others, and make life feel worth inhabiting.

When a Person Feels Unnecessary

One of the darkest possibilities in the AI age is that many people may begin to feel unnecessary. This feeling may arise even when they are not starving, even when they still possess comfort, and even when they remain outwardly functional. The deeper pain is not always material. It is existential. A person wonders whether there is still any serious need for them.

This is dangerous because human beings can endure hardship more easily than they can endure the belief that they no longer matter. A person who feels needed can bear difficulty. A person who feels unnecessary may begin to come apart even in relative comfort.

That is why the meaning crisis cannot be treated as a soft or secondary issue. It reaches into mental health, civic energy, family stability, and the long-term moral vitality of a society.

The Crisis of the Self in an Age of Convenience

Modern societies already show signs of this problem. Many people are overstimulated, under-formed, and inwardly tired. They have access to information, entertainment, and convenience on a scale no earlier generation possessed, yet many still feel restless, fragmented, and unsure of what gives life coherence.

Artificial intelligence could either deepen or correct this problem. It could deepen it if it turns life further toward passivity, artificial companionship, frictionless entertainment, and avoidance of difficult growth. It could help correct it if it frees human beings from lower burdens while pushing them toward higher forms of learning, service, care, creativity, and wisdom.

That is why the decisive issue is not whether AI creates ease. It is what kind of self that ease encourages.

Why This Is Not a Romantic Defense of Suffering

To warn against a meaningless life is not to romanticize suffering. There is no nobility in avoidable misery, pointless drudgery, or crushing insecurity. If artificial intelligence can reduce some of those burdens, that is a real good. It can free time, lower cost, extend capability, and widen access in important ways.

But freedom from misery is not the same as a meaningful life. The argument of this chapter is not that people should remain burdened in order to stay serious. It is that, once lower burdens are reduced, human beings must still rise toward higher purposes. If they do not, emptiness may simply replace hardship.

A wise civilization therefore tries to do two things at once: reduce unnecessary suffering and cultivate lives worthy of the freedom that results.

The Need for a Higher Reorientation

This means the answer to the meaning crisis cannot be purely economic. Even a fairer distribution of AI-driven wealth would not be enough by itself. People would still need structure, commitment, forms of service, reasons for discipline, and a sense that their lives participate in something larger than consumption.

The AI age therefore forces a reorientation. Human beings may need to move from identities organized mainly around necessity and output toward lives organized more consciously around vocation, care, teaching, mentorship, service, creativity, wisdom, and legacy.

That shift is difficult, but it may also be one of the great opportunities of the age. If we are wise, AI may not only expose the danger of a meaningless life. It may also push us toward a deeper understanding of what a meaningful life truly is.

The danger of the AI age is not only that machines may do more. It is that human beings may do less with their own freedom. A civilization that gains convenience but loses purpose will not flourish, no matter how advanced its systems become.

Chapter 17

If Not Work, Then What?

Horizon Insight: A meaningful life cannot rest on one job title alone. The AI age calls for a broader architecture of work, family, service, learning, community, creativity, and vocation.

Beyond the Job-Centered Life

A Full Life Is Built Before It Is Needed

One practical lesson is that people should not wait until a job disappears, a retirement begins, or a crisis arrives before building a broader life. Family, friendships, service, learning, health, music, faith, craft, civic life, and community cannot be manufactured instantly at the moment of loss. They must be cultivated before they are urgently needed.

In that sense, the AI age gives us a warning while there is still time. Do not let one role carry the entire weight of identity. Build a life with several strong pillars. Then, if one pillar changes, the whole house does not collapse.

This is one of the practical lessons of the AI age. A person whose whole identity rests on one professional title is more fragile than a person whose life has several serious pillars. The answer is not to care less about work. It is to let work stand inside a wider human order so that usefulness, love, service, and learning do not disappear when one economic role changes.

This chapter should not be read as an argument against work. Work will continue to matter. But the AI age may ask us to build lives broad enough that one job title does not carry the whole weight of meaning.

If artificial intelligence changes the structure of work deeply enough, then one of the most practical questions of the future becomes this: if work no longer remains the sole organizer of life, then what should take its place? This question is not abstract. It goes directly to the shape of ordinary human existence.

For generations, modern life has often been built around a job-centered model. Education prepared people for work. Adulthood was measured by work. Identity was often tied to work. Social respect was often granted through work. Time itself was structured by work. Even leisure was often defined as recovery from work.

That model may now come under pressure. Not because work disappears entirely, but because its old centrality may weaken. If that happens, human beings will need more than economic adaptation. They will need a broader architecture for living.

Why This Question Matters So Much

A society can survive many kinds of change if it still knows how to form serious lives. But if the older framework weakens and nothing stronger replaces it, then people may gain flexibility while losing shape. They may gain time while losing direction. They may gain assistance while losing the disciplines that once held life together.

That is why this chapter matters. It asks not merely how people can remain employed, but how they can remain fully human if employment no longer carries the same total organizing power. The right answer cannot be passive entertainment, endless consumption, or the hope that comfort will somehow become meaningful by itself.

A serious life still needs structure. It still needs responsibility. It still needs effort directed toward something worth doing.

The Need for a Broader Life Architecture

If work becomes less central, then other dimensions of life must become more intentionally developed. A meaningful life may need to be built more consciously out of several pillars rather than one dominant pillar alone. These may include paid work, but also family, friendship, caregiving, service, learning, craft, community, health, citizenship, spiritual or moral development, and forms of contribution not reducible to wages.

This does not mean work stops mattering. It means work must be placed inside a larger human frame. The old danger was making work too small and mechanical. The new danger is making life too empty if work weakens. The answer is not withdrawal. The answer is a richer architecture of living.

A person with such an architecture is less likely to collapse when one role changes. They still experience themselves as responsible, connected, and called toward something worthwhile across more than one domain of life.

Portfolio Life Versus Drift

One possible future is what might be called a portfolio life. This does not mean a shallow collection of hobbies. It means a life structured around multiple serious forms of responsibility and contribution rather than a single narrow work identity. A person may combine paid work with caregiving, learning, mentoring, local service, artistic creation, civic involvement, and community building.

The idea is not fragmentation. It is integration. A portfolio life becomes meaningful only if its elements are ordered toward a coherent sense of purpose.

This is very different from drift. Drift is unstructured, reactive, and passive. Portfolio life is intentional. It asks: what are the responsibilities, relationships, and forms of contribution that together make this a serious life? If a society cannot make that distinction clearly, then flexibility may dissolve into shapelessness.

A Life with More Than Work

Freedom becomes healthier when it has places for learning, service, family, health, and reflection.

The image shows that freedom becomes healthier when it has a structure of family, service, learning, health, and reflection.

Vocation Is Larger Than a Job

One word that becomes increasingly useful here is vocation. A job is a role. A vocation is a calling. A job often answers what one does for income. A vocation reaches further. It asks what one is called to contribute across the whole shape of a life.

In an age of AI, vocation may become more important because the older equation between occupation and identity may weaken. A person may work in several different roles across a lifetime, or move between formal employment and other forms of contribution. What gives continuity in such a life cannot simply be one job title. It must be something deeper.

Vocation provides that deeper continuity. It helps a person ask: what kind of good am I here to strengthen? What kinds of people or communities am I called to serve? What patterns of effort and responsibility make my life coherent?

This does not solve every practical problem. But it gives a larger organizing principle, and without some such principle, freedom may become disorienting rather than fruitful.

The Role of Family and Care

Modern market societies often undervalue the human importance of family formation, caregiving, and intergenerational responsibility because these do not always show up clearly in conventional economic metrics. Yet these are among the most foundational forms of human work.

To raise a child well, care for an aging parent, sustain a marriage, support a struggling friend, or hold a household together requires patience, sacrifice, steadiness, and love. These are not secondary to civilization. They are among the conditions that make civilization possible.

If AI reduces some forms of economic pressure, one hopeful possibility is that societies may eventually be able to value these forms of contribution more honestly. A human life organized partly around care is not a lesser life. It may, in many cases, be a deeper one.

Learning as a Lifelong Organizing Principle

Another pillar of a broader life architecture is learning. Not only learning for a credential, and not only learning to remain employable, but learning as part of what it means to remain mentally alive. In a fast-changing world, the person who continues learning is not merely adapting to markets. They are continuing to grow as a human being.

This matters because one danger of a post-work imagination is stagnation. If people imagine freedom mainly as relief from effort, they may become less alive over time. But learning keeps the mind in motion. It renews curiosity, builds humility, and sustains the sense that life remains unfinished in a fruitful way.

That is why education and lifelong learning are not side topics in the AI age. They are among the main supports of a meaningful life beyond narrow economic function.

Community and Local Contribution

A good life also needs community. People do not flourish alone. They need to belong somewhere concrete: a neighborhood, a group, a place of service, a local network of friendship and mutual responsibility. Modern societies often weaken these bonds by making life highly mobile, highly individualized, and heavily organized around market demands.

If work loses some of its old centrality, that may be dangerous unless community becomes stronger. But it may also be an opportunity. People may have more room to invest in the human worlds immediately around them rather than living entirely under the pressure of institutional productivity.

Local contribution matters because it gives shape to freedom. A person who belongs somewhere and contributes somewhere is less likely to become lost in abstraction or consumption alone.

Freedom Requires Form

One of the great illusions of modern life is that freedom means lack of constraint. In reality, freedom without form often becomes emptiness. A meaningful life is not built only by removing necessity. It is built by choosing patterns worthy of devotion.

That is why the future cannot be imagined simply as liberation from work. Liberation from necessity may create room, but room itself must be filled. It must be filled by responsibility, relationships, discipline, service, care, craft, and commitment to goods larger than appetite.

Freedom therefore requires form. The AI age may make that truth easier to see because the weakening of old structures will force many people to build new ones more consciously.

A Serious Life in the Age of AI

A serious life in the age of AI is not one in which technology disappears. It is one in which technology is placed in service to a larger human order. The person still works, but does not reduce the whole self to work. The person still values productivity, but not at the expense of purpose. The person still uses intelligent tools, but does not allow those tools to define the meaning of life.

In such a life, work becomes one important strand among others: family, service, learning, community, moral growth, creativity, stewardship, friendship, and perhaps spiritual depth. These strands do not compete against one another if well ordered. They strengthen one another.

That may be one of the great tasks of the years ahead: not merely to survive technological change, but to build lives that are broad enough, deep enough, and ordered enough to remain meaningful within it. Then the horizon widens again from the interior life to the physical world: health, care, mobility, homes, robotics, and independence.

That is where the next part of the book begins.

Chapter 18

From a Survival Society to a Meaning Society

Horizon Insight: A meaning society will not emerge automatically from abundance. It must be built through dignity, contribution, care, wisdom, and fair distribution.

This chapter is the society-level synthesis of the previous meaning chapters. The earlier chapters asked how an individual can remain needed, purposeful, and serious. This chapter asks what kind of civilization would make such lives easier to build.

A Civilizational Possibility

A Meaning Society Must Be Practiced Locally

A meaning society will not begin as an abstract national slogan. It will begin in smaller places: a grandparent teaching a child to ask better questions, a retired person mentoring a young worker, a family caring for an aging parent, a local group helping adults learn new tools, a school asking students to explain their thinking instead of merely submitting polished output.

This is why I return again and again to families and communities. Large systems matter, but culture is lived locally. If AI gives us more capability, the question is whether ordinary people use that capability to become more isolated and entertained, or more generous, skilled, connected, and wise.

A meaning society will not be created only by national policy or technological abundance. It will be practiced locally first. It will appear when a retired person mentors a young worker, when a school values curiosity over shortcuts, when a company trains beginners instead of only automating them away, when a family uses AI to support conversation rather than replace it, and when a community gives older adults a way to remain useful.

The phrase may sound large, but its first evidence will be small. A society becomes more meaningful when people are given real ways to contribute, care, learn, build, teach, and belong. AI may create some material conditions for that possibility. But human beings must still choose the culture that makes it real.

If artificial intelligence helps reduce some forms of scarcity, lowers the cost of certain goods and services, improves health, supports caregiving, and weakens the older bond between labor and survival, then humanity may eventually face a civilizational possibility unlike the one that shaped most of history. We may begin moving, however unevenly, from a survival society toward a meaning society.

That phrase should be handled carefully. It does not mean that difficulty disappears, that tragedy vanishes, or that human life becomes effortless. It does mean that the organizing center of society may gradually shift. For most of history, survival came first. Food, shelter, labor, physical risk, and the struggle against scarcity dominated the structure of life. Meaning existed, of course, but it existed under the pressure of necessity.

If AI changes the material structure of civilization deeply enough, necessity may no longer dominate in the same way for everyone. That possibility raises one of the largest questions in the whole book: if survival is no longer the only organizing principle, what then becomes the purpose of human life?

What a Survival Society Is

A survival society is one in which large portions of human life are organized around the struggle to secure basic conditions of existence. People work because they must. Institutions are designed heavily around material production, economic necessity, and the management of scarcity. Identity is closely tied to usefulness because usefulness is tied closely to survival.

In such a society, there is often little room for asking larger questions except at the margins. Meaning, beauty, philosophy, contemplation, art, mentorship, and inward development may exist, but they are often treated as luxuries compared with the pressure of getting through the day, feeding a family, holding a job, or protecting one's place in the world.

This is not a criticism of earlier humanity. It is simply a recognition of how strongly necessity has shaped civilization. Much of what we call normal life was built under the rule of scarcity.

What Changes if Scarcity Eases

If artificial intelligence, combined with other technological changes, gradually reduces certain forms of scarcity, then some of the structures built around survival may begin to loosen. The cost of producing certain services may fall. Some healthcare interventions may become more precise and effective. Transportation may become easier. Support systems may grow smarter. Productivity may rise while the required human labor for some forms of output falls.

None of this guarantees justice. It does not automatically ensure broad distribution. But if it happens on sufficient scale, it could widen the practical room in which human beings live. When survival pressure eases, the human question changes. One is no longer asking only, "How do I get through?" One is increasingly forced to ask, "What is all this freedom for?"

That is a more difficult question than it first appears. Necessity can be harsh, but it also gives life a built-in structure. When necessity weakens, meaning must become more intentional.

The Opportunity of a Meaning Society

A meaning society would be one in which human beings are increasingly called to organize life around contribution, vocation, relationship, wisdom, service, beauty, and inner development rather than survival alone. In such a society, the highest human achievements may no longer be measured mainly by how efficiently people produce under pressure, but by what kind of human world they help create.

This could be one of the great hopes of the AI age. If the machine increasingly carries more of the lower material burden, then the human being may have more room to rise toward what is higher: raising children with seriousness, caring for others well, studying deeply, creating beauty, strengthening community, cultivating wisdom, mentoring the young, preserving truth, and building institutions that help life become more humane.

That would not mean productivity no longer matters. It would mean that productivity is no longer the sole or highest horizon of civilization.

From a Survival Society to a Meaning Society

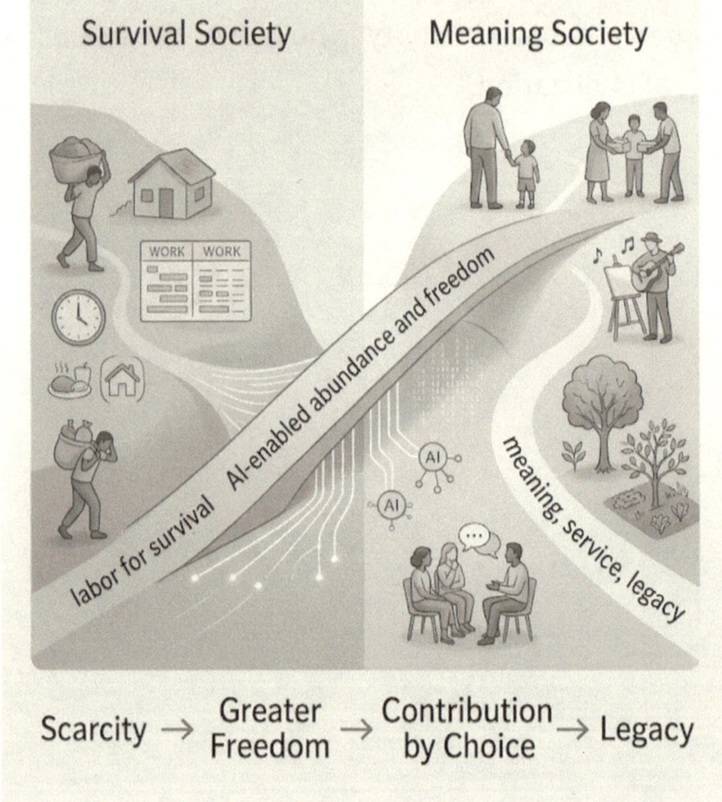

As survival pressure eases, the human task may shift from necessity toward meaning.

The movement from survival pressure to purpose is not automatic; it requires deliberate choices, stronger institutions, and deeper human character.

The Legacy Era

One way to describe this possible transition is to call it the Legacy Era. In a survival society, the central question is often how to secure one's place. In a meaning society, the central question may increasingly become what one leaves behind.

Legacy here should not be understood only in financial terms. It includes whom one shaped, what one protected, what standards one upheld, what courage one modeled, what institutions one strengthened, what truths one preserved, and what kinds of people one helped bring into being.

That is why the future may not reduce the importance of human beings. It may, in a deeper sense, intensify it. The old measures of output may weaken, but the moral and relational measures of a life may become more visible.

The Danger Inside the Opportunity

But this transition contains danger as well as hope. A society relieved of some necessity does not automatically become wise. It may instead become passive. If people are given more freedom without stronger formation, they may drift into distraction, consumption, and inward weakness. A meaning society cannot be built on abundance alone. It must be built on character, culture, and institutions that still call people upward.

That is why the movement from survival society to meaning society is not guaranteed to be an ascent. It could become a decline if ease replaces seriousness. Abundance without purpose may produce not flourishing, but a softer form of emptiness.

That is one reason the earlier chapters on meaning, education, and the human response matter so much. A civilization with powerful tools and weak people may become richer in convenience while becoming poorer in soul.

The Distribution Question Still Matters

It is also important to say plainly that a meaning society cannot emerge for everyone if abundance is captured only by a narrow elite. If AI-driven productivity is concentrated too heavily in a small number of firms, countries, or powerful groups, then many people may experience not liberation from scarcity, but exclusion from a richer future shaped by others.

That is why the question of distribution remains essential. The move toward meaning society depends not only on technology, but on whether enough people have real access to security, time, education, and room for higher forms of life. Without that, the phrase becomes rhetorical. A society cannot ask people to pursue meaning if it still keeps them trapped in desperation.

So, this chapter does not cancel the earlier economic chapters. It depends on them. The possibility of a meaning society exists only if dignity and participation are protected materially as well as spiritually.

A Different Standard of Success

If this transition unfolds, societies may eventually need a different standard of success. Today, success is often measured heavily by growth, output, wealth, and status markers tied to productive advantage. Those things will not vanish, but they may no longer be sufficient to describe a flourishing civilization.

A more mature standard of success would ask harder questions. Are people becoming more capable of care? Are they more free in the meaningful sense, not just the consumer sense? Are communities stronger? Are families healthier? Are older adults more respected? Are children better formed? Are people using increased capability to deepen human life, or simply to accelerate distraction?

These are not sentimental questions. They may become some of the most practical civilizational questions of the AI age.

Why This Chapter Belongs in This Book

This chapter belongs here because the meaning question cannot remain private. A person can choose service, learning, family, and vocation, but society can either support those choices or make them harder. Education, housing, caregiving, local community, public health, transportation, taxation, and technology policy all shape whether people have room to build meaningful lives.

The AI age therefore requires more than personal adaptation. It requires social design. A meaning society is not a fantasy of leisure. It is a civilization that uses abundance to strengthen contribution, care, wisdom, enterprise, family, community, and human dignity.

The expanding frontier of AI thus leads to a larger possibility: not only a more technologically capable society, but a society forced to ask what human life is really for when necessity loosens its grip.

This reflective section prepares the final chapter of the book, where the horizon, the practical response, and the meaning question come together into one civilizational choice.

Part IV Summary

Work, Meaning, and a New Architecture of Contribution

Part IV argued that the AI age forces us to rethink work as more than employment. Work has long provided income, structure, competence, recognition, and the feeling of being needed. If AI changes the old labor system, human beings will still need contribution, responsibility, community, and a reason to get up in the morning. The task ahead is not merely to replace jobs with leisure, but to build a society where meaning, dignity, service, family, enterprise, and human depth can continue to grow.

Part V Introduction

How Every Generation Can Help Shape Civilization

The book now turns from shared human principles to family action. The older generation, the working generation, and the young generation each need a different kind of courage: the courage to stay engaged, to translate experience upward, and to build depth before speed.

Chapter 19

Advice for the Older Generation in the Age of AI

Horizon Insight: Older adults do not have to become fast to remain valuable. Their gift is wisdom, steadiness, memory, mentorship, and the courage to stay in the conversation.

When I speak to older adults, I often say this directly: you do not need to become an AI engineer, but you should not remove yourself from the AI conversation. You have children, grandchildren, friends, finances, health questions, community responsibilities, and a lifetime of wisdom. AI will touch all of those areas.

Reframing Your Role in a Changing World

When discussions about artificial intelligence arise, attention often turns immediately toward younger generations or those still building their careers. People ask what students should study, what skills workers should develop, and which professions may rise or fall. These are important questions, but they can create a misleading impression: that the older generation stands somehow outside the transformation, watching it happen from the sidelines. That is not true.

The older generation is not outside this moment. It holds a uniquely important place within it. In fact, one of the great mistakes society may make in the age of AI is to assume that value belongs only to speed, novelty, and technical fluency. If we make that mistake, we misunderstand both technology and human life.

Many older adults feel a quiet concern beneath the surface of the technology conversation: the fear that the world is moving ahead without them. That feeling is understandable. But it is not the whole truth. Artificial intelligence may replace certain tasks, but it cannot replace perspective. In fact, as the world becomes more complex, fast-moving, and sometimes confusing, the ability to interpret events thoughtfully becomes more valuable, not less.

I have seen this issue very clearly in my AI lectures. After the talk, some retired people come up to me and say, "I am glad I am already retired. I do not have to worry about AI and jobs anymore." I always understand the feeling behind that sentence. They are relieved that they do not have to compete in a changing labor market.

But I also believe that sentence reveals a dangerous misunderstanding. The AI age is not only about jobs. It is about remaining capable, independent, respected, and connected. A retired person may not need to protect a career in the same way a younger worker does, but that does not mean AI can be ignored. If older adults stop learning entirely, they may become technologically obsolete in daily life. They may find it harder to understand documents, protect themselves from scams, use new services, communicate with younger generations, or participate in important family conversations.

Even more importantly, they may lose the ability to guide the young. A grandparent or retired professional may have wisdom, judgment, patience, and perspective, but if younger people feel that this older adult has no understanding of the world they are entering, they may stop listening. That would be a great loss. The older generation does not need to become expert in AI, but it does need enough familiarity to remain credible, engaged, and useful.

Older adults do not need to outrun AI; they need to stay present and engaged.

From Experience to Wisdom

We should distinguish between experience and wisdom. Experience is what one has lived through. Wisdom is what one has learned from those experiences and how one interprets them.

Artificial intelligence can process enormous amounts of information. It can retrieve patterns, summarize documents, answer questions, and generate plausible recommendations at extraordinary speed. But it does not possess lived experience. It does not carry emotional memory in the human sense. It does not understand consequences as a human being understands them, because it has not suffered, hoped, regretted, loved, failed, persevered, or watched events unfold across decades of life.

Over time, people accumulate perspective. They begin to see patterns in human behavior. They recognize trade-offs. They become less easily dazzled by novelty and more alert to unintended consequences. They understand that a good answer in theory can fail in life if it ignores context, timing, character, or human motivation.

In conversations with younger people, I have often noticed that what they need most is not more information. Usually, they are already surrounded by information. What they need is better interpretation. They need help seeing what matters, what is noise, what is temporary, and what is worth building a life around.

This is where wisdom becomes indispensable. Your role is not to compete with machines on speed. It is not to prove that you can process more information than an AI system. Your role is to offer clarity in a world that is becoming increasingly complex.

The Older Generation as Guide

Experience becomes guidance when it helps the young ask better questions.

Experience becomes valuable when it helps the young ask better questions.

AI as Chief of Staff

For older adults, the phrase "chief of staff" is useful because it keeps the relationship in the right order. A chief of staff organizes, reminds, drafts, summarizes, and prepares. But the chief of staff does not become the president. The older adult remains the person with memory, judgment, conscience, family ties, and final responsibility.

Used this way, AI can help older adults remain more independent. It can reduce friction without taking over the self. The practical goal is not to become technical for its own sake. The goal is to preserve agency: to understand enough, verify enough, and participate enough that technology serves dignity rather than quietly replacing confidence.

For the older generation, one of the most useful ways to think about AI is not as a rival, but as a kind of chief of staff. A chief of staff does not replace your mind. It helps you organize, remember, compare, draft, and navigate complexity more efficiently. That is exactly where AI can be especially valuable.

An older adult can use AI to summarize long documents, explain unfamiliar language, compare options, draft letters, prepare questions for a doctor, interpret contracts at a first-pass level, organize travel plans, generate checklists, clarify financial terms, and help sort through confusing choices. For many people, these are not trivial conveniences. They are tools of practical independence.

Used well, AI can support memory, reduce cognitive overload, and help people remain more fully engaged with the world around them. Seen this way, it can function not as a threat to dignity, but as an ally of dignity.

AI as a Support for Later Life

AI can reduce overload when people remain thoughtful users.

AI can be utilized as a chief of staff to help day-to-day tasks.

A Practical Checklist for Older Adults

[ ] **Practice One Useful Tool:** Use AI for one clarifying task, such as explaining a letter, preparing doctor questions, or summarizing a document.

[ ] **Verify Urgent Messages:** Call back through a trusted number before sending money, sharing information, or reacting to fear.

[ ] **Join the Family Conversation:** Ask younger relatives what they are using, and share your concerns without embarrassment.

[ ] **Use AI as Chief of Staff:** Let it organize and clarify, but keep your judgment, conscience, and relationships in charge.

Learn Enough to Stay in the Conversation

There is, however, another reason older adults should learn at least enough AI to be conversant with it. In many families and communities, younger generations often assume that older adults are technologically distant. When that assumption hardens, something important can be lost. The older generation may still possess deep wisdom, but it may no longer be heard as readily if it is seen as entirely outside the world that younger people inhabit.

For that reason learning enough to stay in the conversation matters. It is not about pretending to be a technical expert. It is about earning enough credibility so that younger generations no longer see you as a technological fossil, but as someone still engaged, still growing, and still capable of understanding the world they are living in.

Once that credibility is earned, something important becomes possible. Older adults are more likely to be heard when they speak about judgment, character, patience, responsibility, and the larger meaning of life. Seen this way, learning AI can help preserve not only independence, but influence.

Acceptance Creates Responsibility

This earned acceptance matters for another reason as well. Older adults do not merely want to be accepted by younger generations for social comfort. They need that acceptance in order to fulfill one of their responsibilities: helping guide the young in preparing for the age of AI.

Younger people may understand new tools faster, but they do not automatically understand what kind of life those tools should serve. That is where the older generation still has something essential to offer. It can warn against shallowness, remind the young that usefulness is not the same as worth, and help them see that wisdom, discipline, and humanity matter even more in a high-tech era.

So, learning enough AI to stay in the conversation is not a superficial matter. It is part of preserving a moral voice within the family and across generations. Acceptance becomes the doorway through which guidance can still be given.

Another group of people comes to me with almost the opposite emotion. Some are grandparents. Some are parents in the middle of life, carrying work responsibilities while worrying deeply about their children. They say, "I feel helpless. My children and grandchildren are facing an uncertain future, and I do not know how to help." My answer is always gentle but firm: you can help more than you think.

You do not have to predict exactly which jobs will disappear or which tools will dominate ten years from now. You can help the young by asking better questions. Are they using AI to learn, or only to shortcut? Are they building attention, character, communication, and judgment? Are they becoming more curious or more passive? Are they learning how to work with people, not only with machines? Are they developing the human skills that become more important when technology becomes more powerful?

This is where older adults still matter deeply. They may not move as quickly with the newest tools, but they can help younger people keep life in proportion. They can remind them that speed is not wisdom, output is not character, and convenience is not meaning. That kind of guidance may become even more valuable in the AI age.

Do Not Let the Tool Replace the Self

Yet there is an important caution here. A supportive tool can become a weakening tool if it replaces rather than strengthens the self. The goal is not to stop thinking and let the machine do all interpretation. The goal is to use the machine to clear noise so that one's own judgment can operate more effectively.

This means older adults should use AI as an aid to clarity, not a substitute for conscience. Ask the machine to explain, compare, and summarize — but still decide with your own mind. Let it help you ask better questions, but do not give it the final authority over what you trust.

The difference matters because the dignity of the older generation is not preserved by dependence. It is preserved by strengthened agency.

Wisdom Becomes More Visible in a Fast World

One of the great errors of youth-centered cultures is to confuse novelty with wisdom. Speed matters in some domains, but speed alone does not tell a person what deserves to be built, trusted, or resisted. The older generation often carries something more valuable than novelty: proportion.

Proportion means the ability to see events in context, to distinguish passing noise from lasting importance, to recognize repeating patterns, and to resist panic when others are swept up in extremes. A person who has lived through multiple technological waves, political cycles, market manias, and human disappointments often has a steadier sense of scale.

That steadiness may become more important in the AI age. A society intoxicated by capability may need people who can ask slower questions: What kind of world are we becoming? What are we forgetting? What must not be traded away for convenience? Older adults may be uniquely positioned to ask those questions with moral seriousness.

Your Value Is Not Speed

This point deserves to be said plainly. The value of the older generation is not primarily speed. It is not how quickly one adopts every new interface or masters every new system. The deeper value is wisdom, emotional steadiness, pattern recognition, mentorship, and the ability to judge life beyond the excitement of the present.

Modern culture often underprices these qualities because they are harder to measure than technical output. But they remain essential to families, communities, and institutions. A machine can retrieve information. It cannot replace a life lived long enough to understand what matters after novelty fades.

That is why older adults should resist the temptation to measure themselves by the standards least suited to their deepest strengths.

Mentorship as a Generational Role

The older generation also has an important role in mentorship. Young people and middle-aged adults alike are navigating a world of rapid change, uncertainty, and shifting structures. They need more than technical advice. They need models of seriousness, encouragement, perspective, and moral confidence.

A mentor does not need to know every tool in detail to matter. Often what people most need is not another expert in features, but someone who can help them see life in proportion, stay grounded under pressure, and remember that usefulness is not the same as worth.

When I sing for older adults in senior communities, I am reminded again and again that later life is not only a stage of decline. I often see people respond with warmth, memory, humor, and emotional presence in ways that reveal how much depth is still there. That experience has made me even more convinced that the older generation has something essential to give, not only in affection and encouragement, but in steadiness and lived human truth.

This is especially moving with residents living with dementia or serious memory loss. Sometimes a person who can no longer carry on an ordinary conversation will suddenly respond when a familiar song begins. A face softens. A hand taps. A few words return. The room becomes more human. Those moments remind me that people are not only thinking machines. We are memory, feeling, music, recognition, and connection. AI may imitate a voice or generate a song, but it does not share the lived meaning of a song remembered across a lifetime.

This is one reason the older generation remains deeply relevant. It can help steady the generations behind it, not by pretending the old world is unchanged, but by reminding them of what remains humanly true even when tools change rapidly.

Keep Learning Without Panic

None of these means the older generation should ignore AI. On the contrary, it should keep learning — but without panic. Learn enough to use helpful tools. Learn enough to protect yourself from deception. Learn enough to stay intellectually alive and practically capable. But do not mistake frantic adaptation for wisdom.

A calm learner often learns better than a frightened one. The goal is not to become a younger person again. The goal is to remain fully yourself while continuing to grow. Curiosity is still available in later life, and perhaps even more fruitful when joined to maturity.

This is another form of dignity. To keep learning without surrendering to one's center is one of the strongest responses an older adult can make to a rapidly changing world.

The Opportunity of This Stage of Life

In some ways, older adulthood may carry a special opportunity in the AI age. A person may be less trapped than younger generations by the demand to build identity entirely through work. They may have more room to think about legacy, mentorship, service, family, community, art, reflection, and the wider meaning of life.

This does not mean all older adults have ease. Many face health issues, financial pressure, loneliness, caregiving burdens, or grief. But it does mean that this stage of life may offer a perspective from which the larger human questions can be seen with greater clarity. And those larger questions may become increasingly important precisely because AI is so powerful.

The older generation therefore has a role not only in adapting to the future, but in helping society ask what the future should be for.

The advice for the older generation is therefore several things at once. Use AI as support, not as a source of intimidation. Learn enough to remain credible to younger generations. Use that credibility to help guide them in preparing for the age of AI. And do not underestimate the continuing importance of wisdom, sovereignty, steadiness, and mentorship in a time of rapid change.

Chapter 20

Advice for the Middle Generation in the Age of AI

Horizon Insight: The middle generation must adapt without sacrificing family. Audit tasks, learn AI, strengthen human advantages, and build portable value.

I have great sympathy for the middle generation. Many are carrying children, aging parents, mortgages, careers, health worries, and leadership responsibilities at the same time. They cannot simply stop life and reinvent themselves from the beginning. They need practical ways to build on what they already know while preparing for the changes ahead.

The Generation Under the Greatest Pressure

Among all generations, those in the middle stage of life often carry the heaviest load during periods of major change. They are not only trying to preserve their own relevance. They are usually supporting children, helping aging parents, managing mortgages or major financial obligations, and holding together the practical structure of family life. When artificial intelligence begins to reshape work, they do not face that change in isolation. They face it while carrying responsibility in multiple directions at once.

That is what makes the middle generation's position uniquely demanding: they must maintain current workflow productivity using new systems while simultaneously acting as the primary emotional and strategic anchor for the households that rely on them. Younger people may have more time to experiment. Older people may already be beyond the most career-intensive stage of life. But the middle generation must adapt while continuing to perform, often changing under pressure while others still depend on them.

For that reason, the middle generation needs advice that is realistic, calm, and strategic rather than dramatic. Panic is not helpful. Denial is not helpful either. The right response is disciplined adaptation — learning how to use AI in ways that increase leverage without losing judgment, human connection, or long-term direction.

I heard this anxiety directly from parents after my AI lectures. Some do not ask first about their own jobs. They ask about their children. They say, in effect, "I do not know what future my children are entering, and I do not know how to help them prepare." That concern is deeply human. My advice is not that parents must predict every future job or master every AI tool. Their first responsibility is to stay calm, keep learning, and help their children build the durable human strengths that will matter in any future: curiosity, attention, character, communication, judgment, and the ability to use AI without becoming dependent on it.

The Middle Generation
at the Crossroads

Working adults carry career pressure
while guiding children and aging parents.

Working adults carry career pressure while guiding children and aging parents.

Avoiding the Two Worst Reactions

When major technological change arrives, two bad reactions often appear. The first is panic. People imagine that everything they have built is about to collapse, that their experience no longer matters, and that they are already too late. The second is denial. People assume their role is too established, too relational, or too specialized to be affected at all.

Both reactions are dangerous. Panic leads to confusion, poor decision-making, and emotional exhaustion. Denial delays adaptation until the environment has already changed around you. The better path is to recognize something more subtle. Your experience still matters, but it matters differently. You are no longer protected simply because you have years in the field. You are protected when you can turn those years into judgment, leadership, trust, human connection, and skill in directing new tools.The goal, then, is not to throw away your experience. It is to translate it upward. A person who has been in the workforce for twenty years should not be trying to compete with AI only on speed or routine output. They should be learning how to combine field knowledge, practical wisdom, and human relationships with new forms of machine leverage.

Translate Experience Upward

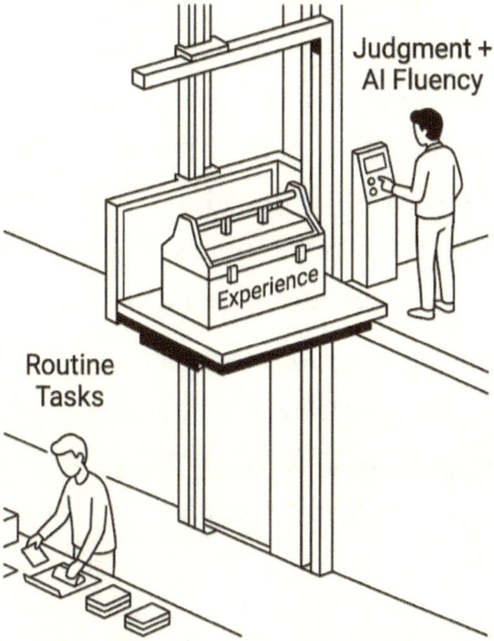

Experience becomes stronger when lifted into judgment and AI fluency.

The goal is not to throw away experience, but to add AI fluency to it.

Understand Your Work as a Bundle of Tasks

One of the most practical things a middle-aged professional can do is examine work honestly. A job is not one thing. It is a bundle of tasks. Some tasks are repetitive, structured, and document-heavy. Others depend on trust, negotiation, interpretation, judgment, and managing people in messy circumstances.

This distinction matters because AI rarely replaces an entire profession all at once. It more often compresses parts of the role. Drafting, standard analysis, summarization, routine reporting, data organization, and first-pass preparation may increasingly be handled with machine assistance. That does not automatically make the human being less valuable. But it does force a choice. Either one's role shrinks with the automatable parts, or one moves more deliberately into the higher-value layers that remain.

The right question is therefore not simply, "Can AI affect my job?" The right question is, "Which parts of my work are becoming cheaper, and which parts of my work become more important precisely because the lower layers are becoming cheaper?"

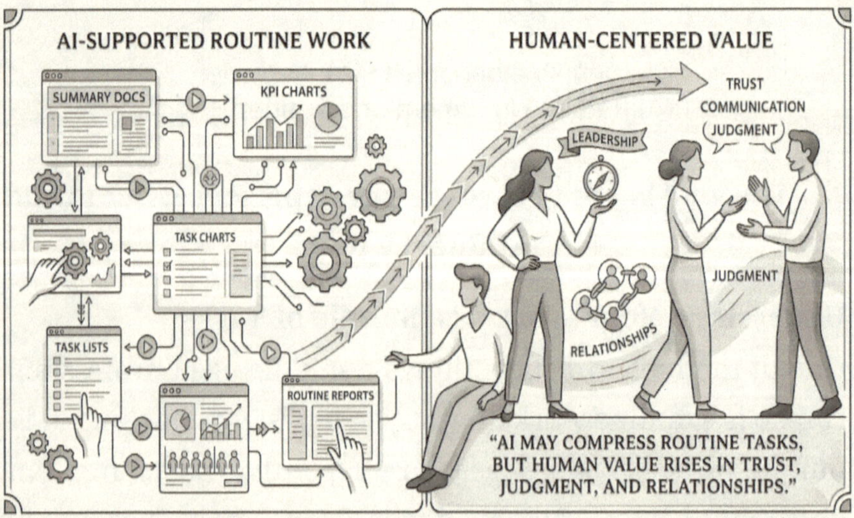

A job is safer when the human part moves toward judgment, trust, and responsibility.

Human Interaction as a Strategic Advantage

The most durable advantage in the age of AI often lies in human interaction — building trust, communicating clearly, aligning teams, influencing outcomes, and carrying responsibility in ways other people can rely on.

In my discussions, I often summarize this shift simply: AI will not take your job. People who use AI well will. That statement is not perfect in every literal case, but it captures an important truth. Those who use AI to handle routine tasks can redirect their effort toward activities that require judgment and connection. They move from executing tasks to shaping direction.

Individuals who act as bridges across departments, who translate ideas across functions, and who build trust across boundaries become increasingly difficult to replace. In many organizations, the new advantage will not lie only in being technically competent. It will lie in being the person who can connect technical capability to human reality.

The Emergence of the Solo Corporation

One of the more striking possibilities of the AI age is what might be called the emergence of the solo corporation. A single capable person, equipped with strong tools, may increasingly be able to perform work that once required a small team: drafting material, analyzing information, building presentations, comparing options, producing prototypes, and coordinating basic processes.

This does not mean every professional should quit immediately and become independent. It does mean, however, that middle-aged adults should begin thinking differently about leverage. A person who understands both their field and new tools may be able to create side income, build advisory work, launch niche services, or strengthen negotiating power within an organization.

The larger lesson is that value may become more portable. Instead of depending entirely on one institutional ladder, a person may gradually build capabilities that travel with them.

The Profession Plus One Strategy

A useful way to think about this is the profession plus one strategy. Keep your real profession. Keep your domain depth. But add one adjacent layer of leverage that makes you more powerful in the AI age.

For one person, that plus one may be AI fluency inside an existing field. For another, it may be stronger communication and presentation ability. For someone else, it may be product thinking, client development, data interpretation, leadership, or the ability to build a small independent practice beside formal employment.

The point is not random diversification. It is strategic reinforcement. The middle generation often cannot afford reckless reinvention. But it can often afford intelligent extension.

The Thirty-Minute Rule

One practical discipline I would recommend is very simple: a thirty-minute rule. Spend at least thirty focused minutes a day learning, testing, or integrating AI-related tools and workflows that connect to your actual life and work.

Thirty minutes may sound modest, but over time it compounds. It keeps panic from taking over because it turns vague fear into concrete movement. It also prevents the opposite mistake, which is endlessly reading about change without building any practical familiarity with it.

The goal is not to become a full-time technologist. The goal is to stay in motion. In periods of transition, consistent modest adaptation is often more powerful than bursts of frantic reaction.

Adapt Upward, Not Out of Fear

The better response is steady movement toward higher human value.

The middle generation does not need panic or denial; it needs disciplined movement toward higher human value.

Do Not Sacrifice Family to Adaptation

There is also a warning here. The middle generation is especially vulnerable to turning adaptation into another source of frenzy. A person may respond to technological pressure by working longer, chasing every new tool, and becoming even more unavailable to family life than before.

That would be a mistake. The middle generation is not carrying only a career. It is often carrying a family system. Children are watching how adults respond to change. Aging parents may be depending on practical steadiness. A spouse may be sharing the strain. If adaptation comes at the cost of becoming less human at home, then something important has gone wrong.

For that reason, the middle generation must think not only about relevance, but about translation. It often serves as the translating generation — translating a fast-changing future into something livable and responsible for others who depend on it.

Portable Value Over Institutional Illusions

Many middle-aged adults were formed in a world where loyalty to one institution appeared to offer relative stability. That model is weaker than it once was. Institutions still matter, but they may no longer provide the same long-term guarantees that earlier generations assumed.

This means a wiser mindset is to build portable value. Portable value includes trusted relationships, a reputation for reliability, communication skill, judgment, domain depth, and the ability to use new tools intelligently. These things do not belong only to your employer. They travel with you.

In the AI age, the middle generation may need to think less like occupants of fixed ladders and more like builders of resilient capability.

Steadiness Is Part of the Gift

The middle generation also has something important to offer that goes beyond adaptation itself. It often becomes the stabilizing center of families and institutions. Younger people may bring experimentation. Older people may bring long memory and perspective. The middle generation often has to turn those into workable daily reality.

That is why emotional steadiness matters so much. A panicked middle generation can spread fear downward and upward at the same time. But a calm, learning, realistic middle generation can help both children and parents navigate the transition with less confusion.

This is one reason the middle generation's role is larger than personal survival. It is helping carry continuity through change.

A Practical Checklist for the Middle Generation

[ ] Audit Your Task Bundle: Separate routine tasks from judgment, relationship, and leadership tasks.

[ ] Build Profession Plus One: Keep professional depth, then add AI fluency that strengthens your field directly.

[ ] Apply the Thirty-Minute Rule: Spend focused weekly time testing one real workflow without surrendering final judgment.

[ ] **Protect Family Presence:** Do not let adaptation pressure consume the relationships you are trying to protect.

The advice for the middle generation is therefore clear. Do not panic. Do not deny. Examine your work honestly. Move your value upward. Add AI fluency without losing human depth. Build portable leverage. Learn steadily. And remember that your task is not only to survive the transition, but to translate it responsibly for others who depend on you.

Chapter 21

Advice for the Young Generation in the Age of AI

Horizon Insight: Young people should not become weaker versions of the machine. They need depth, attention, character, proof of real work, and the courage to solve messy-world problems.

If I were speaking to a grandchild or a student, I would put the advice this way: do not use AI to avoid becoming strong. Use it to become stronger. Let it explain, challenge, quiz, and expand your thinking, but do not let it replace the slow inner work of becoming a capable human being.

You Are Growing Up Inside the Change

I have special concern for the young because they are not merely adapting to a tool introduced late in life. They are being formed inside the tool's atmosphere. The habits they build now - attention or distraction, honesty or shortcutting, curiosity or passivity - may shape not only their school performance, but the kind of adults they become.

The great temptation for young people is to use AI to avoid the very struggle that builds strength. But the great opportunity is just as real. AI can become a tutor, simulator, debate partner, language coach, coding helper, research assistant, and creative companion. The difference depends on whether the young person remains the learner, builder, and moral owner of the work.

The younger generation enters the age of artificial intelligence in a very different position from everyone else. Older generations remember a world before the internet became central. The middle generation remembers a time before smartphones and algorithmic platforms were woven into ordinary life. But many young people are growing up in a world where digital systems, instant answers, and machine-generated content already feel normal.

That creates a special challenge. If technology has always been present, convenience can be mistaken for understanding, and fluency can be mistaken for wisdom. Young people may assume that because answers are available, learning itself is less important. Passive dependence can grow quietly before one has even developed a strong mind of one's own.

That is why the younger generation cannot approach AI casually. Your challenge is not merely to use the tool. Your challenge is to become the kind of human being the tool cannot replace. That requires more than technical skill. It requires depth, discipline, and a serious way of building a life.

Young people should use AI to deepen ability, not avoid the work that forms them.

A Practical Checklist for Young People

[ ] **Protect Attention:** Set regular blocks for reading, thinking, practice, study, and real conversation.

[ ] **Use AI to Learn:** Ask for hints, examples, critique, and quizzes before accepting final answers.

[ ] **Build Proof of Work:** Keep drafts, projects, version history, presentations, and explanations that show real understanding.

[ ] **Choose Messy Problems:** Move toward work that combines technology with judgment, care, ambiguity, and real-world complexity.

[ ] **Study Humanity:** Read history, literature, philosophy, art, music, and world affairs so technical fluency has depth.

Do Not Become a Weaker Version of the Machine

One of the greatest risks for young people is that they may gradually become a weaker version of the machine. They may outsource too much thinking, rely too quickly on generated answers, lose the habit of sustained attention, and confuse speed with understanding. On the surface, they may look productive. Underneath, they may be becoming thinner.

That path is dangerous because the machine will usually be better at machine-like behavior. If a young person trains themselves mainly to produce fast outputs, summarize shallowly, imitate fluency, and move from prompt to prompt without deep internal wrestling, then they are competing on the terrain least favorable to their deepest humanity.

The better path is the opposite: become more deeply human. Use the machine, but do not let the machine shape the limits of your mind or your character.

Protect Your Attention

Attention may become one of the most important assets of the young generation. A person who cannot sustain attention cannot build deep understanding, cannot form strong judgment, cannot do serious creative work, and cannot grow through difficulty. Yet modern technology already pressures attention badly, and AI may intensify that pressure by making answers, stimulation, and content even more immediate.

That is why protecting attention is not a small self-help suggestion. It is a life strategy. The young person who can still read deeply, think slowly, wrestle with complexity, and remain present without constant digital fragmentation may gain an enormous long-term advantage.

Attention is not only a study skill. It is a moral and intellectual power. It determines whether you become capable of depth or merely reactive to whatever is placed in front of you.

Endless scrolling, algorithmic stimulation, fragmented attention, and constant comparison can quietly shape the inner life in ways that make serious adulthood harder. The AI age may intensify this if it makes digital environments even more persuasive and personalized. That is why resisting passive digital formation is one of the central life tasks of the young generation. The question is not only what tools you use. It is what tools are using you.

To build a strong life, you will have to reclaim enough of your mind to think, choose, read, study, and love without constant algorithmic intrusion.

In the AI age, protected attention becomes a foundation for depth.

Attention is the private space where judgment, love, and deep thought grow.

Major Plus One

One of the strongest strategies for young people in the AI era is what I would call major plus one. Stay serious about your major. Learn your discipline deeply. If you study law, learn law seriously. If you study medicine, learn medicine seriously. If you study engineering, business, biology, design, or any other field, build real depth in it.

But do not stop there. Add one more layer: meaningful AI fluency. Learn how intelligent tools are changing your field. Learn how to work with them without becoming mentally dependent on them. Learn how to use them to deepen your understanding rather than to avoid the work of understanding.

This is the real meaning of major plus one. It is not superficial multitasking. It is the creation of a bridge identity:

someone who understands both a real domain and the machine systems that are transforming that domain. The bridge person is likely to be very valuable in the years ahead because the future belongs not simply to technical specialists, and not simply to domain traditionalists, but to those who can connect both worlds intelligently.

The key warning, however, is this: do not become a shallow operator of tools. If you use AI in a lazy way, you may appear efficient while your understanding grows weaker. That is a dangerous illusion. Use AI so that your grasp of the subject becomes stronger, more precise, and more flexible, not thinner.

Choose Messy-World Problems

Young people are often told to chase whatever looks most advanced, most digital, or most fashionable. But one of the wiser strategies may be to move toward messy-world problems. AI is very powerful in structured environments, with large amounts of data and relatively clear optimization goals. It is still much less naturally at home in the full complexity of the physical, human, and social world.

This means there may be unusual value in fields and roles that require human contact, judgment under ambiguity, practical care, embodied work, community leadership, teaching, negotiation, entrepreneurship, design in real environments, and forms of responsibility that do not reduce cleanly to screen-based optimization.

The lesson is not to avoid technology. It is to place yourself where technology and real human complexity meet. That is often where opportunity will be greatest.

Be Deeply Educated in Humanity

Another dimension is often overlooked. It is not enough to be technically trained. In the age of AI, young people need to be deeply educated as a human being.

This means engaging seriously with history, philosophy, literature, art, music, world affairs, culture, and the long human conversation across generations. These fields do more than deliver information. They help you understand how human beings think, suffer, hope, fail, adapt, and create meaning.

In my experience teaching college students, I have often observed a concerning pattern. Many focus narrowly on their major, aiming only to pass exams and graduate. Over time, they become technically trained but intellectually limited. They struggle to write well, read deeply, and engage meaningfully in conversation.

In the age of AI, this becomes a significant disadvantage. Machines may become extraordinarily capable in processing and generating information, but they do not provide wisdom, moral judgment, or depth.

That is why I believe young people should resist becoming narrowly trained task performers. They need to read well, write well, speak well, present confidently, understand history and world affairs, and develop the human skills that allow them to work with others. In my business life, I saw the same lesson from another direction: people with strong human skills usually advanced farther because they could build trust, bridge departments, handle customers, and carry responsibility. AI makes this lesson more urgent. The more machines can do, the more young people need the human depth to direct them wisely.

AI may become the brain. But humans must remain the heart.

Those who develop both technical capability and human depth will be far better prepared to navigate the future.

In a technological age, young people need to be deeply educated in humanity.

A strong major becomes more valuable when paired with a second human or technical edge.

Using AI to Learn, Not to Skip

One of the greatest risks is using AI to bypass effort rather than enhance learning. Students can now generate essays, solve problems, and complete assignments in minutes. But speed is not the goal. Understanding is.

If AI is used as a shortcut, it weakens the ability to think independently. If it is used as a partner, it strengthens understanding. A more effective approach is to use AI to explore ideas, challenge assumptions, generate examples, test explanations, and identify gaps in your thinking.

The goal is not to finish faster. The goal is to think more deeply. Do hard thinking yourself. Let the machine help you stretch understanding, not replace the internal struggle by which understanding becomes your own.

AI should help students think more deeply, not skip the work of understanding.

Students should use AI as a mentor.

In the years ahead, proof of work will matter more. People will increasingly want to know not only what you produced, but whether you can explain it, defend it, improve it, and use it in the real world. Surface fluency will not be enough.

Build Human Skills, Not Only Technical Skills

At the same time, a serious challenge is emerging. Many young people are spending increasing amounts of time on screens — social media, gaming, messaging, short-form content, and fragmented digital interaction — while spending less time in real conversation. The result is not just distraction. It can become a gap in human capability.

This gap is not mainly technical. In fact, many young people are extremely fluent with devices. The problem is different. Some of the most valuable skills in the age of AI will be speaking well, listening deeply, reading people well, comforting others, persuading thoughtfully, building trust, and working productively with other human beings. These are not skills that develop mainly through passive screen interaction.

If you cannot communicate well, if you cannot read emotion, if you cannot build real relationships, then technical fluency alone will not make you effective. Technology amplifies what you can do, but human connection often determines the impact of what you do. In many situations, what matters deeply is not whether you have access to information, but whether people trust you, understand you, and want to work with you.

That is why human skills are becoming more important in the AI era. The machine can produce content. It cannot easily create genuine trust.

Human Skills in a Machine Age

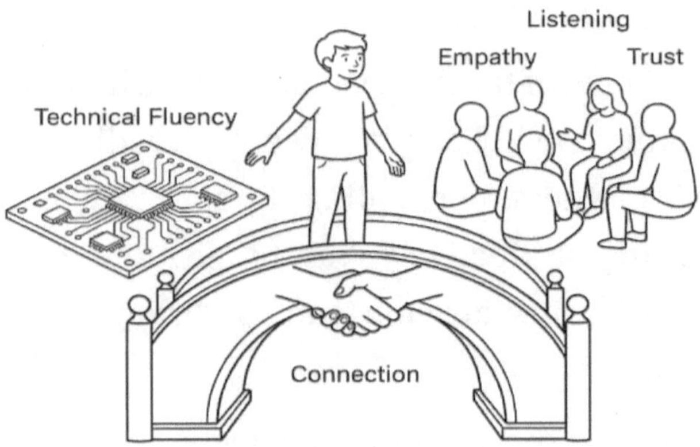

Technical fluency becomes stronger when joined with empathy, communication, and trust.

AI fluency is meaningful only when it strengthens human values.

That is the human edge the young should build: not merely faster output, but the ability to earn trust, explain clearly, and carry responsibility in the real world.

Build Character Before Chasing Efficiency

There is another temptation the young generation will have to resist: the temptation to become efficient before becoming strong. AI can make many things easier, but ease is not the same as maturity. Character is built through responsibility, perseverance, honesty, restraint, courage, and the willingness to do difficult things even when shortcuts are available.

A young person who learns to protect attention, tell the truth, carry responsibility, remain teachable, and stay steady under pressure will often be building something more valuable than technical fluency alone. Technical fluency can be learned quickly. Character usually cannot.

In the AI age, one of the greatest advantages may simply be to become the kind of person other people can trust with complexity, ambiguity, and responsibility.

The Hope for the Young Generation

The young generation also carries immense possibility. It may become the first generation able to use intelligent systems from the beginning of adult life while still choosing depth, humanity, and seriousness. If it can do that, it may become extraordinarily capable.

A young person who protects attention, studies something real, uses AI wisely, builds character, works on messy-world problems, and understands humanity deeply may have access to tools more powerful than any previous generation while still remaining anchored in what matters deeply.

That combination could be remarkable. But it will not happen automatically. It will require intention, resistance to shortcut culture, and the willingness to become strong before becoming merely efficient.

The advice for the young generation is therefore not to fear AI, nor to surrender to it. It is to use it while becoming stronger than it can make you on its own. Protect attention. Build real depth. Add tool fluency. Do hard thinking yourself. Study humanity. Build character. Seek proof of work. Resist passive digital formation.

Part V Summary

How Every Generation Can Help Shape Civilization

Part V translated the book's argument into family action: older adults should stay engaged and protected, working adults should translate experience upward, and young people should build depth, character, and proof of real learning before speed. Together these generational responses turn the book's horizon into a shared family and civic responsibility.

Chapter 22

Final Thoughts — The Role of Humans in the Next Civilization

Horizon Insight: The future is not simply what machines become capable of doing. It is what human beings choose to protect, strengthen, and become in response.

A Rare Moment in History

A final chapter should not repeat the whole book, but it should gather its spirit. The spirit of this book is not anti-technology. I have spent much of my life inside technology, teaching it, building it, selling it, and explaining it. But long experience has made me less impressed by power alone and more interested in what power serves.

That is the question I want the reader to carry beyond the final page. AI will become more capable. That much seems clear. The more difficult question is whether we will become more capable of wisdom, care, courage, judgment, and love.

We arrive here not with fear, but with seriousness. Every generation eventually faces a question larger than its own comfort. For us, the question is how to keep the age of intelligent machines deeply human.

We are living through one of those rare moments in history when the structure of society begins to shift beneath our feet. Most generations experience change gradually. The world they inherit resembles the one their parents knew, and they assume that their children will live in something broadly similar. The details change, but the deeper structure remains recognizable.

But occasionally history accelerates. The Agricultural Revolution reshaped how we lived. The Industrial Revolution reshaped how we worked. The Computer Revolution reshaped how we stored, processed, and exchanged information.

Today we are entering another phase — one that is not just about tools, machines, or information systems, but about intelligence itself. That is what makes this transition so unusual. It does not simply extend what we can do. It changes what we may no longer need to do and therefore forces us to reconsider what remains essential to being human.

It forces us to reconsider what remains essential: our judgment, our capacity for deep relationship, and our willingness to carry responsibility for the choices we make. The ultimate victory in the age of AI will not belong to the most efficient algorithm, but to the most mature humanity. As you close this first volume and look toward the horizon of the next twenty-five years, remember that the future is not something that happens to us: it is a civilization we are actively building at our own kitchen tables, in our classrooms, and in the quiet, daily choices we make to stay in the driver's seat.

Every Generation on the Same Frontier

Throughout this book, we have seen that artificial intelligence affects each generation differently. For the young, it is the environment they are growing up in. For the middle generation, it is a force that challenges stability while also opening new forms of leverage. For the older generation, it is a tool that can extend independence while clarifying the enduring value of experience and wisdom.

Yet despite these differences, we share something important. We are all standing on the same frontier. There is no previous generation that has fully navigated this transition for us. There is no complete map. There is no settled script.

That means our role is not only to adapt individually. It is also to help shape the path forward together. The choices we make — in education, in work, in family life, in institutions, and in public life — will influence how this transition unfolds for those who come after us.

What Really Changes in the Age of AI

For generations, human values have often been tied closely to what we can produce: our ability to remember, calculate, organize, repeat, and execute tasks reliably. In earlier eras this made sense. Those capacities were scarce, and societies depended heavily on them.

Artificial intelligence is beginning to take over more and more of those functions. This does not mean that human beings become less important. It means that the basis of human value begins to shift.

As machines increasingly handle lower forms of cognitive labor, what rises in importance are the capacities that cannot be reduced so easily to output alone: judgment, character, wisdom, responsibility, care, emotional steadiness, interpretation, moral courage, and the ability to hold human life together under conditions of uncertainty.

Seen this way, the AI age does not erase the human role. It exposes it with greater clarity.

The Human Core Must Rise

This may be the deepest argument of the whole book. If the machine becomes more powerful, then the human core must rise rather than shrink. We cannot answer the age of AI by becoming more hurried, more superficial, more dependent, or more passive. That would only turn us into weaker versions of the system we created.

The better response is the opposite. We must become more serious where seriousness is needed, more educated where judgment is required, more human where human presence matters, and more deliberate where technological power tempts us into thoughtless speed.

That is why so many of the chapters in this book converged on the same themes. Education matters more. Character matters more. Meaning matters more. Intergenerational guidance matters more. Truth matters more. Attention matters more. The future is not saved by cleverness alone. It is steadied by stronger people.

As machine power expands, the human center must deepen rather than dissolve.

As AI power expands, the human character must be strengthened.

The Choice Before Us

The age of AI does not present only one future. It presents a field of possibilities. One future leads toward convenience without depth, efficiency without dignity, stimulation without meaning, and capability without wisdom. Another future leads toward wider freedom, deeper education, stronger care, more intentional lives, and a civilization that uses its tools without surrendering to them.

The difference between these futures will not be decided by the machine alone. It will be decided by what human beings honor. Do we honor truth, or only persuasion? Do we honor education, or only credentials? Do we honor wisdom, or only speed? Do we honor care, or only measurable output? Do we honor human formation, or only technical performance?

These are not abstract philosophical luxuries. They are the practical moral choices that will shape the next civilization.

Do Not Measure Human Worth Too Low

One of the great dangers of the AI age is that societies may continue measuring human worth too low. If we continue to define a person mainly by routine productivity, narrow economic function, or surface technical performance, then we will misread both the threat and the opportunity of artificial intelligence.

The opportunity is that the machine may take over some lower burdens and force us to ask again what human beings are really for. The threat is that we may answer that question badly by drifting into passivity, distraction, and shallow dependence.

That is why the central human task is not merely to remain useful. It is to remain worthy of the freedom that more powerful systems may create. A society that gains capability but loses seriousness will not flourish. A society that gains capability and rises in humanity may.

What I Hope We Remember

If there is one thing I hope the reader will remember from this book, it is this: artificial intelligence will change many things, but it does not relieve human beings of the responsibility to become wise.

The machine may help us think faster. It cannot tell us what is worth thinking about. It may help us generate more. It cannot tell us what is worth creating. It may help us manage more complexity. It cannot tell us what kind of life is worthy of a human being.

Those questions remain ours. They belong to parents and children, teachers and students, workers and leaders, citizens and communities, older adults and the young. They belong to every generation because they are finally questions about what sort of civilization we want to inhabit and pass on.

Stay Human

So, my final message is not to fear artificial intelligence, and not to worship it either. Learn it. Use it. Understand it. Let it help reduce confusion, drudgery, and unnecessary burden. Let it become a tool of support, discovery, care, and wider freedom where it can.

But do not forget the things that make life worth living. Think deeply. Keep learning. Protect attention. Build character. Care for people. Strengthen families. Tell the truth. Make something meaningful. Use power responsibly. Help the next generation become strong. Remain open to wisdom wherever life still offers it.

Stay curious. Stay kind. Stay responsible. Stay human.

A Final Sentence

If I had to summarize this whole book in one sentence, it would be this:

Artificial intelligence will change many things, but what matters deeply in life will still be decided by humans — by our wisdom, our values, and how we treat one another. The choice remains ours.

Overall Summary of This Book

This book has argued that AI is not merely a technical event, but a civilizational turning point. It changes tools, work, learning, family life, social trust, and the meaning of human contribution.

The central answer is the same throughout: the machine may become the engine, but the human being must remain the compass.

A humane AI age will require practical skills, educational reform, intergenerational guidance, shared dignity, and a higher view of what human beings are for.

Understanding the Whole Journey: The Four-Book Family Map

This book stands alone as Volume 1, your wide-angle map of the AI age. However, because the challenges of this transition are generational, I have designed a four-volume guide to ensure every member of your family, team, or community has the tools they need to thrive.

The Roadmap for Your Family

As you finish this volume, you may find yourself thinking of others in your life who are standing at different points on the same road. This series is curated as a **Complete Family Collection** to provide a shared compass for everyone you care about.

- The Human Role in the Age of AI, Vol. 1: The Horizon: Guide for Every Generation Facing the Next 25 Years – Your wide-angle map of AI, work, education, meaning, risk, dignity, and the human future.
- The Human Role in the Age of AI, Vol. 2: The Middle Generation: How Professionals Can Adapt, Lead, and Support Their Families
- The Human Role in the Age of AI, Vol. 3: The Young Generation: How Students Can Learn, Think, and Build a Meaningful Life
- The Human Role in the Age of AI, Vol. 4: The Guiding Generation: How Parents, Grandparents, Teachers, and Mentors Can Help the Young Thrive

Who Needs the Other Books?

- **For the professional:** If you have colleagues or family members managing workplace anxiety, Volume 2 is

- their essential professional compass.
- **For the student:** If you have children, grandchildren, or students who need to build "uncopyable" skills and real-world judgment, Volume 3 is their student field guide.
- **For the mentor:** If you are a parent, teacher, or community leader looking for actionable ways to support the young generation, Volume 4 provides the practical framework for adult guidance.

Why the Collection Matters

The age of AI is not only a school story or a workplace story; it is a family story. If you have benefited from this roadmap, consider sharing the complete four-volume collection with a loved one. By sharing the full series, you can help every generation in your family gain the tools needed to navigate this transition together with confidence.

About the Author

Dr. Harry Wong has spent his life at the intersection of artificial intelligence, business, teaching, family, and the arts. He earned his PhD in Computer Science with a focus on artificial intelligence from the University of Toronto, one of the important birthplaces of modern AI, and has written more than fifty research papers.

He has taught at the University of Toronto and San Francisco State University, built and led an enterprise software company for thirty-five years, and served major corporate customers, including many Fortune 500 companies that purchased products from his company for optimization, communication, and migration to the enterprise cloud. Those experiences shaped his conviction that technology alone is never enough; judgment, communication, trust, and human understanding determine whether powerful tools truly serve people.

Harry is also a father of four and a grandfather of three. His family life deeply shapes this four-book series. He writes not only as a researcher, teacher, entrepreneur, and grandfather thinking about the world his grandchildren will inherit.

In addition, Harry is a classically trained singer who has performed in opera houses, concert halls, community settings, and senior communities. Singing for older adults, including people living with memory loss, has reinforced one of the central themes of his work: human beings are not only processors of information. We are memory, emotion, relationship, dignity, music, and meaning.

Across all four volumes of The Human Role in the Age of AI, Harry brings these parts of his life together to help readers understand AI clearly while protecting what is most human.

A Note to Readers

Thank you for reading Volume 1 of *The Human Role in the Age of AI*. I wrote this series to help families, professionals, students, parents, grandparents, teachers, mentors, and communities face the AI age with clarity rather than fear.

My goal in writing this four-volume series is to provide a common language for families. We are living through an era of profound change; this series is intended to be a shared compass for parents, grandparents, and students alike. My own experience as a father, grandfather, teacher, entrepreneur, and employer has shaped how I see this responsibility. I hope this book helps you understand the horizon of the AI age, and that it also becomes a family doorway into the rest of the series. The age of AI is not only a technological story; it is a family story. A person who understands the horizon more clearly can help their children, grandchildren, students, and community members enter the future with more steadiness.

Continue the Journey

To share this roadmap with your entire family—or to gift the complete collection to a colleague or loved one—explore the complete four-volume collection. The full series gives each generation a guide suited to its responsibilities and helps families navigate the age of AI with a shared language and greater confidence.

Dr. Harry Wong

www.ingramcontent.com/pod-product-compliance
Lightning Source LLC
LaVergne TN
LVHW100255210926
848480LV00003B/923
* 9 7 9 8 9 9 6 1 3 8 8 0 7 *